THE ILLUSTRATED
RHODODENDRON

R. dalhousiae by J.D.Hooker from *Rhododendrons of the Sikkim Himalaya*

THE ILLUSTRATED RHODODENDRON

Their classification portrayed through the artwork of Curtis's Botanical Magazine

Pat Halliday

Illustrations by

John Curtis, Sydenham Edwards,
Walter Hood Fitch, Joseph Dalton Hooker,
Christabel King, Mary Mendum, Valerie Price,
Rodella Purves, Stella Ross-Craig,
Matilda Smith,Lilian Snelling, Margaret Stones,
Wendy Walsh and Ann Webster

PUBLISHED BY THE ROYAL BOTANIC GARDENS, KEW

First published 2001

Design by Fiona Bradley, page make up by Margaret Newman
Media Resources, Information Services Department,
Royal Botanic Gardens, Kew.

Production Editor: Brian Mathew

ISBN 1 900347 99 7

Origination by Magnet Harlequin

Printed in Great Britain by The Bath Press

CONTENTS

ACKNOWLEDGMENTS

I would like to thank the staff of the Royal Botanic Gardens Kew and Wakehurst Place, the 'Rhododendron Team' at the Royal Botanic Gardens Edinburgh, and the staff of the botanic garden at Glasnevin, who in their several ways have helped in the preparation of this book.

My thanks also are extended to Victoria Matthews and Brian Mathew (who succeeded her as editor) — hard-working, long-suffering colleagues who have done so much towards achieving publication. In particular I would like to thank Keith Ferguson for his support and encouragement throughout the project.

Finally, my especial thanks to all those anonymous supporters — enthusiasts of fine botanical artwork, rhododendrons or both — whose existence inspired the creation and development of this book.

INTRODUCTION

The content of a book of this sort, involving personal choice of what is considered to be beautiful, interesting or special in any particular way, is bound to be not only subjective, but also arguable and will probably be judged to be guilty of the crime of omission. Both growers and taxonomists will undoubtedly quibble at some of the included illustrations which, for one reason or another, they will consider inaccurate; but this is primarily a book of 'pretty pictures', given which standard the only acceptable criticism must be of the author's opinion as to which plants qualify for inclusion in the above categories, or can be conveniently dubbed 'pretty'.

At the very beginning of this project, it was decided that every subsection of every section of the genus *Rhododendron* should be represented in the book. Immediately problems surfaced. Not every subsection had been illustrated for the *Botanical Magazine* or the *Kew Magazine*, some plates were too poor or were unavailable but, worst of all, some subsections were not in cultivation so no colour illustrations, traditionally drawn from verified live material, could be prepared. The first two problems were easily solved and new plates were commissioned, but for those species which were not in cultivation and/or of which no live material could be obtained, line drawings had to be substituted; these were prepared from herbarium material and photographs. The number of illustrations representing each subsection was influenced by variation within that group or by my personal preference, or a combination of both criteria. No further significance can be attached to it. However, I am not entirely responsible for every plate chosen. For various reasons, substitutions had to be made; partly because, if a comprehensive range of colour was to be achieved, numerous pink-flowered species had to be disregarded (so many rhododendrons have pink flowers !) and, of course, there were similar problems with plant size.

As well as the new colour plates and four black-and-white drawings, there are other anomalies. Some previously unpublished plates, which for unknown reasons were never written up for the *Botanical Magazine*, although presumably were prepared for that purpose, together with two plates of hybrids by Fitch which were too splendid to exclude and an unfinished plate entitled '*R. argenteum*' [see *R. grande*, Plate 63] by the same artist, which admirably demonstrates his working style, are all included. The two paintings of *R. arboreum* really have no business to be included in a book of plates from the *Botanical Magazine*, but they are relevant in that they illustrate the differing abilities of two Victorians, who have left us such a wonderful legacy — Joseph Hooker, who introduced gardeners to innumerable rhododendrons, and Walter Fitch who was employed by Hooker to paint them and other plants portraits. The two illustrations of *R. arboreum* celebrate their association; the field sketch is by Hooker, the finished painting is Fitch's version of it.

As regards the revision by which the plates are arranged, this too is a purely personal choice. As a taxonomist working at Kew, the Edinburgh revision was the obvious one for me to use. This does not necessarily mean that it is better than other classifications, merely that it was the most convenient system to work with, given that the herbarium material at Kew is arranged according to that revision. For the Vireya rhododendrons I have followed Sleumer's account in *Flora Malesiana* (1966).

These, then, are the main reasons governing the selection of the plates in this book. The task gave me the excuse to spend many pleasurable hours browsing through Kew's extensive collection of original *Botanical Magazine* and *Kew Magazine* paintings, with the prospect of sharing the results of that pleasure with countless other *Rhododendron* enthusiasts. Surely that is a good enough *raison d'être* for the present volume.

Pat Halliday

(1995)

BRIEF HISTORY OF CURTIS'S BOTANICAL MAGAZINE

William Curtis's famous Botanical Magazine made its debut on the 1st of February 1787 and has continued in an unbroken series ever since, right through all the troubled times of the nineteenth and twentieth centuries. It is the oldest botanical, and one of the oldest scientific, periodicals still being published and is thought to be the world's longest surviving magazine in colour. Its founder, William Curtis (1746-1799) trained as an apothecary in London but his real interests lay in natural history and he soon sold his share in the apothecaries business and bought a piece of land in Bermondsey on which to cultivate native plants. In 1773 he was appointed the 'Praefectus Horti and Demonstrator' at the Chelsea garden of the Worshipful Society of Apothecaries of London but left this after about five years and in 1779 started the London Botanic Garden on Lambeth Marsh, between Westminster Bridge and what is now Waterloo Station. Later, the garden was moved to a less marshy position at Brompton and was known as the Brompton Botanic Garden.

Curtis's initial venture into the publication of fine botanical art was the folio-sized Flora Londinensis, a work which depicted the plants growing within a ten-mile radius of London. However, this, although beautifully executed and produced, was not a commercial success, perhaps because of its rather parochial approach in illustrating only local plants, and Curtis suffered a considerable financial setback from the venture. This undoubtedly provided Curtis with an incentive to find something of wider appeal and in 1786 the first plates for his new venture, The Botanical Magazine, were prepared, for publication the following year. In fact, in the preface to the first volume, Curtis tells us that the magazine owed its foundation to the frequent solicitations of various ladies and gentlemen who were subscribers to his Botanic Garden. During the preparation of Flora Londinensis Curtis had become acquainted with a young man from Abergavenny who showed promise as a botanical artist, so Sydenham Edwards was brought to London for further training. This proved to be a great success and Edwards contributed many illustrations to both the Flora and, later, to The Botanical Magazine — in the latter case, some 1700 plates.

The first issue of the Botanical Magazine consisted of three hand-coloured plates and was priced at 1s.; the subsequent parts, each of three plates, were produced monthly. We are told that, in the case of the early volumes, over 3000 copies were printed and most were a complete sell out, so Curtis soon recouped his losses on the Flora Londinensis and was relatively well off when he died at the sadly early age of 53 in 1799.

Curtis's concept was to present colour portraits with accompanying descriptions of the many exotic plants which were entering Britain from all over the world at that time and no doubt his Magazine captured the interest of the wealthy gardening public. Now well into its third century, the 'Bot. Mag.' as it is affectionately known, continues the tradition by publishing a miscellany of new, rare, unusual or attractive plants, illustrated in colour by some of the world's leading botanical illustrators and accompanied by texts which are considerably more detailed than those of the original volumes. Not only is there a formal botanical description of the species but there are discussions about its distribution, habitat, relationship to others, its cultivation requirements, conservation status, history, etc. — in short, as complete a profile of the plant as possible.

In recent years, apart from the miscellaneous selection of plants which are illustrated and published as and when they happen to flower — often at the Royal Botanic Gardens, Kew — there has been a move towards devoting at least one of the quarterly parts to a particular topic. This may be a single genus or group of plants — for example the Vireya rhododendrons in Part 3, 1988 — or family (Part 4, 1997 featuring Leguminosae); sometimes it has a geographical theme (Part 4, 1996 reviewed the flora of the Mascarene Islands), and sometimes marks a special event (e.g. the bicentenary of Glasnevin Botanic Garden in Part 4, 1995). In addition to the 'portrait articles' there are feature articles on a wide range of plant topics, for example a series on the Royal National Parks of Nepal by Tony Schilling. The idea of feature articles was introduced in 1984 by the then Editor Christopher Grey-Wilson and carried on by his successors Victoria Matthews in 1990 and Brian Mathew in 1993.

For the first 160 years of the Botanical Magazine the prints were produced either by engraving or by lithography and then hand-coloured by a team of 'colourists', possibly as many as 200 copies of the same plate per week, working from a specimen plate coloured by the original artist. Looking at different sets of the magazine one sees that, inevitably, there is some variation in the colour of the plates, either as a result of different colourists' work or wear and tear through use or simply ageing processes on library shelves. In 1947 hand-colouring gave way to photographic techniques of reproduction, one of the few breaks with tradition that the venerable magazine has experienced! Although now lacking the personal touch to every print, the originals, and the accompanying black-and-white dissection figures, are as finely executed as ever and are well reproduced, recording the precise details of a wide range of plants of interest to gardeners and botanists alike. In the last 212 years, well over 11,000 illustrations have been published in Curtis's Botanical Magazine, and the original water-colours for many of these illustrations are held in the archives of the Royal Botanic Gardens, Kew.

The genus Rhododendron contains some 850 species native to the northern hemisphere, many of which are widely cultivated — indeed rhododendrons are extremely popular garden subjects. They display a wide range of variability, from dwarf plants suitable for the rock-garden to giants which will happily fit only into a large garden or estate, and they possess flowers in every colour except true blue. With the exception of orchids, there are more plates of rhododendrons in the Botanical Magazine than any other group of plants.

With the publication of this book, Kew has made some of these fine illustrations available to a wider public, each accompanied by a new text updating that originally published in the Botanical Magazine. In almost all cases, the original painting, rather than the published engraving or lithograph, has been reproduced.

The sequence of species illustrated in the book demonstrates the revised classification of Rhododendron that has resulted from studies undertaken at the Royal Botanic Garden, Edinburgh. This classification is now generally accepted by the horticultural world, for example the Royal Horticultural Society and most rhododendron enthusiasts. In the case of a few of the sections where there are no, or unsuitable, illustrations, modern paintings have been produced by artists whose work has been published in Curtis's Botanical Magazine.

THE ILLUSTRATIONS

Alphabetical list of Plates, showing page numbers

Frontispiece: *R. dalhousiae* by J.D.Hooker from *Rhododendrons of the Sikkim Himalaya*

HYBRIDS

THE ILLUSTRATIONS

Numerical list of Plates

Frontispiece: *R. dalhousiae* by J.D.Hooker from

Rhododendrons of the Sikkim Himalaya

CLASSIFICATION OF THE GENUS RHODODENDRON

Subgenus **Rhododendron**

sect. **Vireya** (Bl.) Copeland

subsect. 1 **Pseudovireya** (Clarke) Sleumer. Illustrated: *R. retusum*

subsect. 2 **Siphonovireya** Sleumer. Illustrated: *R. herzogii*

subsect. 3. **Phaeovireya** Sleumer. Illustrated: *R. phaeochitum*, *R. konori*

subsect. 4. **Malayovireya** Sleumer. Illustrated: *R. malayanum*

subsect. 5. **Albovireya** Sleumer. Illustrated: *R. aequabile*

subsect. 6. **Solenovireya** Copeland f. Illustrated: *R. pleianthum*, *R. goodenoughii*, *R. orbiculatum*

subsect. 7 **Vireya**

series i. **Linnaeoidea** Sleumer. Illustrated: *R. disterigmoides*

series ii. **Saxifragoidea** Sleumer. Illustrated: *R. saxifragoides*

series iii. **Taxifolia** Sleumer. Illustrated: *R. taxifolium*

series iv. **Stenophylla** Sleumer. Illustrated: *R. stenophyllum*

series v. **Citrina** Sleumer. Illustrated: *R. citrinum*

series vi. **Buxifolia** Sleumer. Illustrated: *R. burttii*, *R. wrightianum*

series vii. **Javanica** Sleumer. Illustrated: *R. macgregoriae*, *R. zoelleri*, *R. impositum*, *R. brookeanum*

sect. **Rhododendron**

subsect. 1. **Edgeworthia** (Hutch.) Sleumer. Illustrated: *R. edgeworthii*

subsect. 2. **Maddenia** (Hutch.) Sleumer. Illustrated: *R. dalhousiae* & var. *rhabdotum* *R. fletcherianum*, *R. formosum*, *R. veitchianum*

subsect. 3 **Moupinensia** Sleumer. Illustrated: *R. moupinense*

subsect. 4 **Monantha** Cullen. Illustrated: *R. monanthum*

subsect. 5. **Triflora** (Hutch.) Sleumer. Illustrated: *R. yunnanense*, *R. trichanthum*

subsect. 6. **Scabrifolia** (Hutch.) Cullen. Illustrated: *R. racemosum*, *R. spinuliferum*

subsect. 7. **Heliolepida** (Hutch.) Sleumer. Illustrated: *R. rubiginosum*

subsect. 8. **Caroliniana** (Hutch.) Sleumer. Illustrated: *R. minus*

subsect. 9. **Lapponica** (Balf. f.) Sleumer. Illustrated: *R. hippophaeoides*, *R. russatum*

subsect. 10. **Rhododendron**. Illustrated: *R. hirsutum*

subsect. 11. **Rhodorastra** (Maxim.) Cullen. Illustrated: *R. dauricum*

subsect. 12. **Saluenensia** (Hutch.) Sleumer. Illustrated: *R. calostrotum*

subsect. 13. **Fragariflora** Cullen. Illustrated: *R. fragariflorum*

subsect. 14. **Uniflora** (Cowan & David.) Sleumer. Illustrated: *R. uniflorum* var. *imperator*, *R. ludlowii*

subsect. 15. **Cinnabarina** (Hutch.) Sleumer. Illustrated: *R. cinnabarinum*

subsect. 16. **Tephropepla** (Cowan & David.) Sleumer. Illustrated: *R. tephropeplum*

subsect. 17. **Virgata** (Hutch.) Cullen. Illustrated: *R. virgatum*

subsect. 18. **Micrantha** (Hutch.) Sleumer. Illustrated: *R. micranthum*

subsect. 19. **Boothia** (Hutch.) Sleumer. Illustrated: *R. leucaspis*

subsect. 20. **Camelliiflora** (Hutch.) Sleumer. Illustrated: *R. camelliiflorum*

subsect. 21. **Glauca** (Hutch.) Sleumer. Illustrated: *R. glaucophyllum*

subsect. 22. **Campylogyna** (Hutch.) Sleumer. Illustrated: *R. campylogynum*

subsect. 23. **Genesteriana** (Cowan & David.) Sleumer. Illustrated: *R. genesterianum*

subsect. 24. **Lepidota** (Hutch.) Sleumer. Illustrated: *R. lowndesii*

subsect. 25. **Baileya** Sleumer. Illustrated: *R. baileyi*

subsect. 26. **Trichoclada** (Balf. f.) Cullen. Illustrated: *R. trichocladum*, *R. mekongense* var. *melinanthum* subsect. 27. **Afghanica** Cullen. Illustrated: *R. afghanicum*

sect. **Pogonanthum** G. Don. Illustrated: *R. trichostomum*

Subgenus **Hymenanthes** (Bl.) K. Koch

sect. **Ponticum** G. Don

subsect. 1. **Fortunea** Sleumer. Illustrated: *R. calophytum*, *R. decorum* subsp. *diaprepes*, *R. fortunei* & subsp. *discolor*

subsect. 2 **Auriculata** Sleumer. Illustrated: *R. auriculatum*

subsect. 3 **Grandia** Sleumer. Illustrated: *R. grande*, *R. protistum* var. *giganteum*, *R. macabeanum*

subsect. 4. **Falconera** Sleumer. Illustrated: *R. hodgsonii*, *R. falconeri*

subsect. 5. **Williamsiana** Chamberlain. Illustrated: *R. williamsianum*

subsect. 6. **Campylocarpa** Sleumer. Illustrated: *R. campylocarpum*, *R. wardii*

subsect. 7. **Maculifera** Sleumer. Illustrated: *R. longesquamatum*, *R. morii*

subsect. 8. **Selensia** Sleumer. Illustrated: *R. selense* subsp. *dasycladum*

subsect. 9. **Glischra** (Tagg) Chamberlain Illustrated: *R. habrotrichum*, *R. crinigerum*

subsect. 10. **Venatora** Tagg. Illustrated: *R. venator*

subsect. 11. **Irrorata** Sleumer. Illustrated: *R. araiophyllum*, *R. kendrickii*, *R. anthosphaerum*

subsect. 12. **Pontica** Sleumer. Illustrated: *R. yakushimanum*, *R. maximum*, *R. macrophyllum*

subsect. 13. **Argyrophylla** Sleumer. Illustrated: *R. adenopodum*, *R. thayerianum*, *R. ririei*

subsect. 14. **Arborea** Sleumer. Illustrated: *R. arboreum*, *R. niveum*

subsect. 15. **Taliensia** Sleumer. Illustrated: *R. wasonii*, *R. lacteum*

subsect. 16. **Fulva** Sleumer. Illustrated: *R. fulvum*

subsect. 17. **Lanata** Chamberlain. Illustrated: *R. lanatum*

subsect. 18. **Campanulata** Sleumer. Illustrated: *R. wallichii*

subsect. 19. **Griersoniana** [David. ex] Chamberlain. Illustrated: *R. griersonianum*

subsect. 20. **Parishia** Sleumer. Illustrated: *R. kyawi*

subsect. 21. **Barbata** Sleumer. Illustrated: *R. smithii*

subsect. 22. **Neriiflora** Sleumer. Illustrated: *R. sanguineum* var. *haemaleum*, *R. forrestii*, *R. floccigerum*

subsect. 23. **Fulgensia** Chamberlain. Illustrated: *R. fulgens*, *R. sherriffii*

subsect. 24. **Thomsonia** Sleumer. Illustrated: *R. thomsonii*, *R. stewartianum*

Subgenus **Azaleastrum** Planchon

sect. **Azaleastrum**. Illustrated: *R. leptothrium*, *R. ovatum*

sect. **Choniastrum** Franch. Illustrated: *R. moulmainense*

Subgenus **Tsutsusi** (Sweet) Pojarkova

sect. **Tsutsusi**. Illustrated: *R. yedoense* var. *poukhanense*, *R. simsii*

sect. **Brachycalyx** Sweet. Illustrated: *R. reticulatum*

Subgenus **Pentanthera** G. Don

sect. **Pentanthera**

subsect. **Pentanthera**. Illustrated: *R. calendulaceum*, *R. luteum*, *R. occidentale*

subsect. **Sinensia** (Nakai) K. Kron. Illustrated: *R. molle*

sect. **Rhodora** (L.) G. Don. Illustrated: *R. canadense*

sect. **Viscidula** Matsumura & Nakai. Illustrated: *R. nipponicum*

sect. **Sciadorhodion** Rehd. & Wils. Illustrated: *R. albrechtii*

Subgenus **Mumeazalea** (Sleumer) Philipson & Philipson. Illustrated: *R. semibarbatum*

Subgenus **Candidastrum** (Sleumer) Philipson & Philipson. Illustrated: *R. albiflorum*

Subgenus **Therorhodion** (Maxim.) Gray. Illustrated: *R. camtschaticum*

HYBRIDS

subsect. **Fortunea** × subsect. **Arborea**. Illustrated: *R.* × *agastum* (? *R. fortunei* × *R. decorum*)

subsect. **Arborea** × subsect. **Campanulata**. Illustrated: *R.* × *batemanii* (?*R. arboreum* × *R. wallichii*)

subsect. **Pontica** × subsect. **Arborea**. Illustrated: *R.* × *clivianum* (? *R. catawbiense* × *R. arboreum*)

THE CULTIVATION OF RHODODENDRONS

In the texts accompanying the illustrations of the many species described in this book there are cultivation notes and observations about their behaviour in gardens. However, some generalisations can be made about the culture of rhododendrons and for this purpose they are separated into temperate and subtropical species.

TEMPERATE SPECIES — subgenera *Rhododendron, Hymenanthes* and the 'Azaleas'.
In spite of the fact that many rhododendrons grow on limestone in the wild and that some are tolerant of lime in cultivation, an acid well-drained soil is a pre-requisite for most of these plants if they are to succeed. Never plant rhododendrons too deeply; the soil level should only just cover the root-ball. A grafted plant, however, can be planted more deeply, as it will form roots from the union if this is below soil level. The plant will then establish more quickly. On no account should the plants be allowed to dry out in hot sun. This is lethal to them. The compost must always be kept moist as Rhododendron roots lie near the surface and need to remain cool. To accomplish this, an annual mulching with organic matter is recommended. A general fertiliser may be applied in the spring, but as rhododendrons normally grow in poor soil, it is not an essential requirement. Pruning should be restricted to removing dead wood and flowers. Apart from the appearance of the plants, removal of dead wood lets in light and air, while the prevention of seed formation preserves vital energy for use in the production of next year's growth and flower buds. Layering is probably the simplest way of increasing stock of this genus, providing that there is a branch low enough to the ground to be pegged down, otherwise air-layering can be an effective way of acquiring rooted plants. Seed is another suitable method of propagation for this genus, but care must be taken to ensure that the chosen parents do not hybridize with neighbouring species, which most rhododendrons will do very readily! If propagation by cuttings is attempted, these are best taken in July and supplied with 'brisk bottom heat'. They should be rooted in an open compost (e.g. 75% sharp sand, 25% peat) and potted on into a 40% loam, 40% granulated peat, 20% coarse sand mix, when well rooted. An admixture of 20% clean leaf-mould and 20% granulated peat can be substituted for the 40% granulated peat, if preferred.

SUBTROPICAL SPECIES — subgenus *Rhododendron* section *Vireya*.
Members of the section *Vireya* are now quite widely grown, although not many are readily available in Britain. These plants are popular and do very well in New Zealand and in Australia, where a great deal of work with hybridizing is in progress. Much has been published on the subject of Vireyas, and many are the variations on the standard methods advocated for growing them. It is not possible to include them all in the limited space available here, but references are given in the book list on page 268. Interested growers are recommended to consult these and other publications for fuller information. The treatment of the montane tropical species of this group is demanding, but not impossible. Vireyas do not have a winter resting period and the climate in which they grow experiences daily variations in temperature which are more or less constant throughout the year, as are the short days; at night, although the temperature can fall to zero, or occasionally below zero, the cold is of short duration and does no permanent damage. In cultivation *Vireya* rhododendrons require a constant ph level of between 6 and 7, rainwater for watering, summer temperatures between 10' and 20'C (50'-68'F) and some shading during the summer months. Good drainage is essential, and an open root-run is appreciated, although they can be grown successfully in containers. Cuttings can be rooted under mist in a 50% peat, 50% sand mixture. High atmospheric moisture should be maintained

even after the cuttings are well rooted. Established plants do best in raised beds, which afford maximum drainage; they can be grown successfully on dead tree-fern trunks, but these are not always available, or are inconvenient, and then a suitable compost must be found. At Kew the Vireyas are grown in an open mixture of two parts coarse peat and one part coarse sand or grit, to which some charcoal has been added. High atmospheric moisture is maintained at all times.

At the Royal Botanic Garden, Edinburgh, a new compost has been introduced, as the old compost was thought to be too heavy and the results over its first season of use were encouraging. The new compost is made up of 2 parts coarse peat, 1 part fine peat, 1 part super-coarse grade Perlite, grit and magnesian limestone. Both Rouse and Argent note that calcium, in the form of calcium sulphate, should be added to the compost as although Vireyas prefer acid conditions, they also need calcium. It is important that *Vireya* rhododendrons are not over-potted; the larger species may be kept small by careful pinching out and cutting back. When growing Vireyas from seed, it should be borne in mind that the seed is relatively short- lived in comparison to other rhododendrons, so it should be collected as soon as it is ripe. Freshly harvested seed will give better results than stored seed. The seed of these plants is minute and care must be taken when sowing to ensure that it is not covered, although a dusting of dry, sterilized sphagnum moss sieved over the seeds is considered beneficial, and adequate light essential to achieve germination. The containers should never be allowed to dry out, and if these conditions are adhered to, germination should take place in about one month. Seedlings should be encouraged to grow on steadily and given a weekly feed. Control of fungi and algae can be effected by sterilization of the containers and soil, then watering with a half-strength fungicide solution. Although this will significantly affect the germination percentage, it is considered to be the best method.

the RHODODENDRON *PORTRAITS*

RHODODENDRON RETUSUM

Subgenus *Rhododendron*
Section *Vireya* subsection *Pseudovireya*

Messrs Rollisson of Tooting Nursery raised the plant figured here, from seed sent to them by their collector, Henschell. Henschell was following in the footsteps of Blume, who first discovered the species in the mountains of west Java, and it was from that general area that the seed was gathered. *Rhododendron retusum* is a high-altitude plant which also occurs in Sumatra where it was found by Dr Horsfield in 1818, growing in shaded situations at 1000 m above sea-level.

Subsection *Pseudovireya* comprises 25 species, only two of which (*R. vanderbiltianum*, *R. adinophyllum*) occur in Sumatra, most of the subsection being native to New Guinea. Docters van Leeuwen (1927) noted that the flowers of this species are protandrous and are visited by honey birds (*Aethopyga*) and bumble bees, both of which probably assist pollination, but although it has long been acknowledged that birds visiting red tubular-flowered *Vireya* species are probable pollinators, the fact has yet to be proven.

The flowers of this species are in themselves small, but there are many of them and the waxy texture of the bright red corollas glows against the dark foliage. In the fine accompanying plate (tab. 4859) produced by Walter Hood Fitch in 1855, the character of this attractive species is perfectly captured.

Rhododendron retusum (Blume) Benn. in Benn. & Br., Pl. Jav. Rar. 88, t. 20 (1838).
R. retusum Reinw. in Verh. Batav. Genootsch. Kunsten 9: 32 (1823), *nom. nud.*
R. retusum forma *angustifolia* Miq., Fl. Ind. Bot. 2, suppl. 1: 251, 586 (1860).
R. retusum var. *macranthum* Hochr. in Candollea 2: 492 (1925).
Vireya retusa Blume, Bijdr. 856 (1826).
Azalea retusa (Blume) 0. Kuntze, Revis. Gen. Pl. 2: 386, 387 (1891).

Description. A terrestrial shrub or small tree up to 7 m, but more usually 2-4 m in height, erect or occasionally prostrate; young growth densely scaly; internodes short, 4-6(-9) cm long. Leaves coriaceous, in pseudowhorls of 3-6(-7) at the ends of the branches, narrowly to broadly obovate-spathulate, 2-6 cm long, 0.8-2.3 cm wide, base cuneate, apex usually retuse, sometimes obtuse, dark glossy green above, paler, matt and scaly beneath; margin recurved; midrib impressed above, prominent beneath, secondary venation inconspicuous; scales disk-like with narrow, entire marginal zone and impressed centre. Petioles 1-5 mm long. Inflorescence terminal, umbellate, (2-)3-8(-10)-flowered. Outer floral bud-scales subulate-ovate or spathulate, up to 1.3 cm long, 8 mm wide. Bracteoles linear-filiform, up to 6 mm long. Pedicels slender, (0.9-)1.2-1.7(-2) cm long, more or less densely long-hairy, scaly at apex only. Calyx oblique, shallowly 5-lobed, at most 4 mm in diameter, densely scaly externally, lobes long-ciliate. Corolla light red, orange-red or intense scarlet, 1.8-2.5(-3) cm long overall, tubular; lobes erect to spreading, broadly oblong-suborbicular, 4(-6) mm long, 3-6 mm wide; tube straight, 1.1-1.8 cm long, dilated at base, (3-)4-6 mm in diameter, widening upwards to 6-7 mm in diameter at base of lobes, variably scaly and densely long white-hairy externally, glabrous within. Stamens 10, subequal, exceeding the tube; filaments glabrous. Ovary conical, 5-angled, densely scaly with a glabrous basal disk. Style clavate, about 1.7 mm long, lengthening and finally exceeding corolla, scaly at the base. Stigma capitate, 5-lobed. Capsule erect, oblong-fusiform, 1.2-1.5 cm long.

Distribution. Sumatra and Java. In the latter country this species is most commonly found in the west, but does occur on Mt Prahu and Mt Merapi in central Java. In the east it is found on Mt Ardjuno.

Habitat. Sub-alpine forests or Vaccinium scrub, either where the soil is rich in humus or is sandy and/or stony. Also near volcanic craters, where the plants are stunted and small-leaved and the stems produce roots, (1350-)1500-3500 m.

Rhododendron retusum
WALTER HOOD FITCH

RHODODENDRON HERZOGII

Subgenus *Rhododendron*
Section *Vireya* subsection *Siphonovireya*

Subsection *Siphonovireya* comprises seven species. The subsection is characterized by scale type (disk-shaped, subcrenate, narrow margin, thick centre), the narrow, elongated corolla-tube and short spreading corolla-lobes which, at most, attain one quarter of the corolla length. In its native New Guinea *R. herzogii* flowers spasmodically throughout the year, and can grow both epiphytically or terrestrially. *Rhododendron herzogii* is widely grown where Vireyas do well and it deserves the attention, especially as it is small enough for container-growing in a modest-sized glasshouse. The strongly fragrant flowers are borne in a terminal umbel of 5-10, are pure white or have a pinkish tube and appear when the plant is quite young. As with many Vireyas, *R. herzogii* can become leggy if not encouraged to make bushy growth by pinching out. For Vireyas in general, Clyde Smith (1989) recommends that every second new growth is removed, which, with good light and open conditions will usually result in two to four new shoots replacing the one taken out. Picking the flowers for the house also helps to keep the plant compact.

Rouse, Williams and Knox (1988) working with Vireya hybrids in Australia, used the result of a *Vireya-Azalea* cross between *R.* 'Dr Herman Sleumer' and *R. herzogii* as the female parent in a cross with *R. bakeri*. One of the resulting seedlings flowered in November 1987, but the flowers were malformed and sterile.

The illustration reproduced here is new and was painted by Mary Mendum.

Rhododendron herzogii Warb. in Bot. Jahrb. Syst. 16: 15, 25 (1892).
R. carrigtoniae Lane-Poole, For. Res. Papua, 130 (1925), *non* F.v. Mueller (1887).
R. agathodaemonis J.J. Sm. in Nova Guinea 18: 99 (1936), *non* J.J. Sm. (1913).

Description. A terrestrial or less often epiphytic, evergreen, erect shrub up to 2 m in height, but often quite dwarf, with scaly branchlets. Leaves coriaceous, in pseudowhorls of 4-6 at upper nodes, elliptic or obovate-elliptic, (2.5-)4-6(-9) cm long, (1.5-)2-4(-5) cm wide, with cuneate to rounded base and broadly acuminate to obtuse apex, upper surface dark green with caducous scales, lower surface paler yellowish green, densely persistently scaly, scales pale yellow. Petioles (3-)5-8(-12) mm long, scaly. Inflorescence an umbel of 5-10 fragrant flowers. Pedicels erect, stout, (4-)6-8 mm long, densely scaly. Calyx a minute, oblique, undulate rim. Corolla erect, becoming horizontal, pure white or with pinkish tube, salverform; tube straight or slightly curved, 4-5.5(-7) cm long, about 2.5 mm in diameter at base, expanding upwards to 3.5 mm, externally densely scaly, densely hairy and sometimes sparsely scaly within; lobes short, spreading, orbicular-elliptic, externally scaly. Stamens 10, exserted from tube; filaments slender, lower two-thirds hairy. Ovary narrowly cylindrical, tapering abruptly to style, densely scaly. Style exserted from tube, densely scaly, becoming less so in upper part. Stigma green, obconical. Capsule reddish, cylindrical, (2.5-)3-4(-5) cm long, scaly.

Distribution. Papua New Guinea (Main Range, from the Wissel Lakes to the Owen Stanley Range, and on Mt Saruwaged).

Habitat. Cliffs, steep banks and slopes in forest and scrub, sluiced areas and old cultivated land, 1530-2500 m.

Rhododendron herzogii
MARY MENDUM

RHODODENDRON PHAEOCHITUM

Subgenus *Rhododendron*
Section *Vireya* subsection *Phaeovireya*

The first collection of this plant was made by Sir William Macgregor on Mt Musgrave in the Owen Stanley Range, in the Central District of Papua New Guinea. Sir William's expedition in 1889 was undertaken while he was administrator of what was, at that time, a British Crown Colony. He later became the first Lieutenant-Governor of the island.

Rhododendron phaeochitum was one of six rhododendrons brought back by the expedition which were new to science and were described in that same year by Baron von Mueller. The species shows a surprisingly disjunct distribution, apparently only occurring in the above locality, the Eastern Highlands District and in the Telefomin subdistrict of West Sepik District, where it was found as recently as 1975, at an elevation of 2800 m. In the wild it produces flowers from April to December. The species is very free flowering in cultivation, but forms a rather untidy plant. In Australia some of the smaller-leaved, spreading Vireyas are grown in hanging baskets, and it has been suggested that small plants of *R. phaeochitum* might prove suitable for this treatment. The name *phaeochitum* refers to the dense, scurfy, dark-coloured, stellate scales which cover the young leaves, stems and flowers. These scales arise from minute tubercles which persist after the scales themselves have rubbed off and are characteristic of the subsection *Phaeovireya* to which *R. phaeochitum* belongs. The subsection comprises about 41 species, of which *R. beyerinckianum* is the species most closely related to *R. phaeochitum*. Indeed, Sleumer (1966) suggested that the two species might possibly prove to be conspecific. The accepted distinction between them is the presence or absence of pubescence on filaments and disk. In *R. phaeochitum* these are densely hairy, but in *R. beyerinckianum* the filaments are glabrous or only sparsely hairy and the disk is glabrous.

The material from which tab. 766 (new series) was painted was supplied by Mr Geoffrey Gorer in September 1977, but was originally collected in 1968 by Mr Michael Black. The painting is by Margaret Stones and represents one of the forms found in the Eastern Highlands District of the subdistrict of Goroka. Plants from the area are usually pink-flowered, although the original form collected by Sir William MacGregor had deep red corollas.

Rhododendron phaeochitum F. v. Mueller in Trans. Roy. Soc. Victoria, n.s. 1(2): 23 (1889).

Description. Straggling epiphytic or terrestrial, evergreen shrub up to 1 m in height, the whole plant densely rusty-scaly when young, branchlets and leaves scaly and tuberculate, becoming glabrescent. Leaves opposite or in pseudowhorls of 3-5, or clustered near the upper nodes, coriaceous, narrowly to broadly elliptic, oblong or obovate-elliptic, 4-9 (rarely to 14) cm long, 3-4 (rarely to 7) cm wide, base rounded or cuneate, apex rounded and apiculate, entire with slightly revolute margins, dark green above when scales have gone, densely scurfy and rusty-scaly beneath, with scales persistent on midrib and veins; scales dark, stellate, detersile, seated on minute tubercles which persist, making the surface rough to the touch when the scales have fallen. Petioles 0.5-1.5(-1.8) cm long, densely rufous-scaly. Inflorescences terminal, umbellate, 3-6-flowered. Floral bracts membranous, caducous, 1-2 cm long, about 1 cm wide, glabrous except for margins. Bracteoles linear, 1-1.5 cm long, 1 mm wide, with fimbriate-ciliate margins. Pedicels 1.2-2 cm long, densely stellate-scaly. Calyx an undulate or more or less lobed, densely scaly rim, 3-4 mm in diameter. Flowers nodding, corolla pink to deep red, tubular with spreading lobes; tube somewhat curved, widening upwards and oblique at base of lobes, 2-3.5 cm long, 4-5 mm in diameter at base of tube, 8-10 mm in diameter at throat, externally scaly, the scales extending along the median line of each lobe, internally laxly pubescent in lower half; lobes 5, broadly obovate to suborbicular, retuse, 8-11(-15) mm long, 8-12 mm wide. Stamens 10, subequal, usually exserted; filaments up to 5 mm long, pubescent in lower two thirds; anthers about 2.5-3 mm long. Disk more or less densely, minutely white-pubescent. Ovary cylindrical, tapering upwards, about 6 mm long, densely rusty stellate-scaly. Style more or less stout, about 2 cm long, densely stellate-scaly almost to tip, sometimes also laxly hairy. Stigma capitate, obscurely 5-lobed. Capsule and seeds not seen.

Distribution. Papua New Guinea.

Habitat. Ridges, including limestone areas, altitude 2100-2560(-2800) m.

Rhododendron phaeochitum
MARGARET STONES

RHODODENDRON KONORI

Subgenus *Rhododendron*
Section *Vireya* subsection *Phaeovireya*

Of all the Vireya rhododendrons, *R. konori* is one of the most magnificent, producing particularly large, fragrant white to pink flowers. The carnation-like scent becomes stronger at night and is possibly one indication of bat-pollination. Argent (1988) reported finding fallen corollas 15 cm wide in the Southern Highlands of Papua New Guinea, but cultivated plants seldom produce flowers even approaching that size. This fact does not detract from the spectacle of a plant covered with blooms which still appear large by comparison with the size of the plant which bears them and with most other species in the section. At two, or even four, metres in height, the species is rather too large for pot culture in a small glasshouse. However, the flowers are produced when the plant is less than 1 m high and while the plant is that small, it is an ideal subject for container-growing in glasshouse or conservatory where its perfume will be especially appreciated. Peter Cox (1973) commented that growth can be improved by grafting.

The accompanying illustration was painted for the *Kew Magazine* (tab. 107) by Rodella Purves. The material from which it was drawn was introduced in 1973 by Dr R.M. Withers of the Australian Rhododendron Society, and was grown at the Royal Botanic Gardens, Edinburgh. It originated in the Sarawaket mountains north of Lae, Papua New Guinea and was collected above Kasanombe. Some years before this introduction in 1969, the clone 'Eleanor Black', with red-flushed white flowers, received an Award of Merit from the RHS when exhibited by M. Black of Grasmere, Westmorland.

Rhododendron konori Becc., Malesia 1: 200 (1878).
R. toverenae F. v. Mueller, in Victoria Naturalist 1: 101 (1884).
R. devriesianum Koord., Nova Guinea 8: 185 (1909).
R. devriesianum subsp. *astrapiae* Foerster in Feddes Repert. 13: 224 (1914).
R. astrapiae Foerster ex Schlechter in Bot. Jahrb. Syst. 55: 160 (1918).

Description. An epiphytic or terrestrial, evergreen shrub up to 2(-4) m in height with green, rounded, fluted or flattened twigs, initially densely scaly, but scales caducous; foliage-buds often dark purplish red. Leaves coriaceous, dark green above, paler beneath, in pseudowhorls of usually 3-6, sometimes more, or scattered along upper internodes, broadly elliptic-oblong, occasionally obovate, 8-14(-21) cm long, 5-7.5(-9) cm wide, base broadly tapered to rounded, apex usually rounded, occasionally obtusely pointed, margin flat, midrib convex on upper surface in bottom quarter of leaf, prominent beneath almost throughout with 6-12 pairs of lateral veins, upper surface initially scaly, the detersile scales silvery and dendroid, lower surface with more persistent, brown, dendroid scales. Petioles green, not grooved, 1.5-3 cm long, initially brown-scaly, later whitish. Inflorescence an umbel of 4-8 strongly fragrant flowers, held more or less horizontally; floral bud-scales dark reddish purple. Pedicels pale green or flushed with pink or red, 6-15 mm long, silvery dendroid-scaly. Calyx a small, more or less oblique, reddish, angled rim, occasionally with 1 or more acute lobes up to 15 mm long. Corolla fleshy, white to pink or becoming pink when fading, with or without pink markings at the angles of the lobes, funnel-shaped, 8-14(-17) cm long, 9-15 cm wide, externally sparsely scaly; lobes overlapping in the basal quarter; tube fluted. Stamens exserted from throat, usually clustered together on lower side of flower; filaments 5-7 cm long, hairy in basal half; anthers 8-12 mm long. Ovary green, tapering into the style, densely hairy and scaly, the hairs obscuring the silvery scales. Style often pink, 6-7.5(-8.5) cm long, densely hairy and scaly except just beneath stigma where it is glabrous. Stigma green, circular, capitate, lobulate. Capsule more or less fusiform, 6-7(-12) cm long, 1.2-1.6(-1.8) cm wide, irregularly dehiscent, the valves twisting. Style persistent, densely hairy, Seeds (including tails) 6-10 mm long.

Distribution. The main mountain chains of mainland New Guinea, and the islands of New Britain and New Ireland.

Habitat. Margins of montane forests, in the open on burned summits, and on marshy ground, 750-2500 m.

Rhododendron konori
RODELLA PURVES

RHODODENDRON MALAYANUM

Subgenus *Rhododendron*
Section *Vireya* subsection *Malayovireya*

This is one of the better known *Vireya* rhododendrons and one of the easiest to cultivate. It was discovered by Dr W. Jack as far back as the early nineteenth century (circa 1823).

Dr Jack was both a botanist and an author, who worked for the East India Company. He discovered *R. malayanum* in Bencoolen, Sumatra, on the summit of Gunong Bunko (known to Europeans as 'Sugar Loaf'), but it was Thomas Lobb who introduced the species into cultivation, while collecting for the elder Veitch in 1854. Lobb also recorded *R. malayanum* from Borneo. At that time the species was known only from Java and Sumatra so Hooker suspected that Lobb's Bornean record for the species was erroneous, but in later years *R. malayanum* was collected many times in Sabah and Sarawak and it is therefore reasonable to suppose that Lobb's record for Borneo is correct.

The species was figured for *Curtis's Botanical Magazine* in 1873, the accompanying text being compiled by J.D. Hooker. The material from which Fitch drew tab. 6045, was collected in Java, and sent to Kew by Messrs Rollisson of Tooting. According to a letter to J.D. Hooker from Wilbraham Buckley on behalf of Messrs Rollisson, this plant was being grown at that time by an 'Amateur' as a graft onto a 'common sort' of rhododendron, but the nursery was growing 'small plants of it'. In his article, Hooker comments that both *R. tubiflorum* (Java) and *R. celebicum* (Celebes) should be referred to *R. malayanum*, there being no distinctive characters on which to separate them. More recently Sleumer, in his revision of the group for Flora Malesiana (1966), while including *R. tubiflorum* as a synonym, retained *R. celebicum* at specific level, but in a different subsection (*Vireya* series *Javanica*). He split *R. malayanum* itself into five varieties, based on the presence or absence of indumentum on the filaments, scales on the corolla, and whether the inflorescences are terminal, or terminal and axillary. One of these varieties (var. *axillare*) has been further subdivided into three forms, based on leaf shape.

Rhododendron malayanum Jack, Mal. Misc. 2: 17 (1822).
R. tubiflorum Reinw. in Bl. Cat. Gewass. Pl. Buitz., 72 (1823), *nom. nud.*
Vireya tubiflora Blume, Bijdr., 855 (1826).
Azalea tubiflora Blume ex DC., Prodr. 7: 725 (1839), in syn.
R. fuscum Blume, Pl. Jav. Pl. *ined.* t.7c, f.1, (1863-83).
A. malayana (Jack) O. Kuntze, Revis. Gen. Pl. 2: 387 (1891).

Description. A variable erect, straggling, terrestrial or epiphytic shrub or small tree up to 5(-6) m, with branchlets mostly slender, internodes 2-10 cm long, and the whole plant densely scaly, the scales red-brown and of two different sizes. Leaves coriaceous, subopposite or in pseudowhorls of 3-5(-6) at upper nodes, glossy above, red-brown beneath, very variable in shape, from narrowly to broadly lanceolate, elliptic or oblanceolate, (4-)5-10(-16.5) cm long, (1-)1.5-3(-5.5) cm wide, tapering to the petiole, apex acute or more usually acuminate, gland-tipped, margin often somewhat revolute; midrib prominent beneath, other venation inconspicuous. Petioles more or less stout, grooved above, (1-)1.2-1.8(-2) cm long. Inflorescence a terminal umbel of (3-)4-8 nodding flowers. Pedicels curved, 6-9(-15) mm long, pubescent and scaly. Calyx almost obsolete, undulate to 5-lobed, about 3-4 mm in diameter. Corolla waxy, pink to dull red, crimson or purplish, tubular with spreading lobes; lobes broadly obovate to more or less orbicular, 5-7 mm long, 4-5 mm wide; tube usually darker in colour than the lobes, straight or somewhat curved, more or less saccate at the base, (1.5-)1.7-2.2(-3) cm long, 3-5(-8) mm in diameter, usually sparsely scaly externally. Stamens 10, unequal, equalling or slightly exceeding the corolla-tube; filaments glabrous; anthers 1.2-1.6 mm long. Disk scaly but not pubescent. Ovary tapering to style, 4-5 mm long. Style equalling or slightly exceeding stamens in length, scaly at base (occasionally scaly for half its length). Stigma as wide as style. Capsule cylindrical-fusiform, somewhat curved, 2.2-2.6(-2.8) cm long, 3(-4) mm wide. Seeds very narrow, about 5 mm long including tails.

Distribution. *R. malayanum* var. *malayanum* occurs in south Thailand, Sumatra, the Malay Peninsula, west Java, Borneo and the Celebes, where the flowers are consistently larger than those from other regions; var. *pubens* is restricted to the Moluccas, while var. *pilosifilum* is said to occur both in the Moluccas and on Mt Kinabalu in Sabah, Borneo. The two remaining varieties, var. *infrapilosum* and var. *axillare*, (all three forms), are supposedly confined to Borneo. However, Argent et al. (1988) found only one form in Sabah and regard *R. malayanum* as an imperfectly understood species complex. It is therefore possible that some records purporting to be *R. malayanum* may be misidentifications of closely related species.

Habitat. (of the species). Rain forest (epiphytically) subalpine scrub or mossy forest (terrestrially), (140-)1320-1800 (-3000) m.

Rhododendron malayanum
WALTER HOOD FITCH

RHODODENDRON AEQUABILE

Subgenus *Rhododendron*
Section *Vireya* subsection *Albovireya*

Described by J.J. Smith in 1917, *R. aequabile* was found on Mt Singgalang in Sumatra, growing at an altitude of 2800 m. Material was collected from this locality by Dr W. Meijer in 1955 and again in 1957, when he introduced the species into cultivation. It was figured for *Curtis's Botanical Magazine* (tab. 673 new series) in March 1974 when the plant at Kew was in flower. This plant had been raised from Meijer's seed, which was distributed by Dr H. Sleumer of the Rijksherbarium, Leiden in 1957. It was painted by Margaret Stones.

At first glance, *R. malayanum* appears to be the species most closely related to *R. aequabile*, but bearing in mind the importance of the morphology of the scales in section *Vireya*, on closer examination it soon becomes apparent that the two species actually belong to different subsections. *Rhododendron aequabile* has equally sized scales on the under surface of the leaves and belongs to subsection *Albovireya*, while *R. malayanum* with two different sizes of scales, is a member of subsection *Malayovireya*. A closer relative of *R. aequabile* is *R. lampongum*, another Sumatran endemic which has yellow flowers and grows at lower altitudes. Sleumer suspected that both plants will prove to be natural hybrids.

Conditions provided for other Vireyas are suitable for *R. aequabile*, the only deviation being in temperature tolerance. Being a montane species, *R. aequabile* can survive lower temperatures than Vireyas from lower altitudes. D.R. Hunt (1974) reported that in the winter of 1973/74, when temperatures fell below 7°C, this and other montane taxa in cultivation at Kew apparently suffered little or no damage. Propagation is effected by heel or nodal cuttings in summer (in Britain) from mid-June to late July when the wood is half-ripe. The species flowers between January and June in the wild.

Rhododendron aequabile J.J. Sm. in Bull. Jard. Bot. Buitenzorg, ser. 2, 13: 451 (1935).
R. album Ridley in J. Fed. Malay States Mus. 8: 58 (1917), *non* Blume (1823).

Description. A much-branched tree or shrub to 4 m in height; young branches sturdy, more or less striate, densely rufous or dark brown-lepidote; internodes 3.5-16 cm long. Leaves persistent, coriaceous, shiny dark green, in pseudowhorls of 6-9 at the top 2 or 3 nodes, elliptic, (5-)7(-13) cm long including petiole, (1.7-)2.5-5 cm wide, base cuneate, apex subacute to obtuse, sometimes apiculate, scaly but soon glabrescent above, with dark rufous, equally sized, persistent scales beneath, contiguous to overlapping, and with their margins shallowly and irregularly incised, veins inconspicuous, in 6-10 pairs, anastomosing near leaf margin. Petioles grooved above, 0.9-3 cm long. Inflorescence umbellate, 5-12-flowered. Outer floral bud-scales ovate, acuminate, up to 2.5 cm long and 1.0 cm wide, inner bud-scales linear, spathulate, erose at apex, up to 1.5 cm long, about 3 mm wide, glabrous. Bracteoles filiform, up to 1 cm long. Pedicels nodding and more or less arcuate, 1-2 cm long, densely lepidote, the scales contiguous to overlapping. Calyx oblique, undulate to shallowly lobed, (3-)4-5 mm in diameter, densely lepidote. Corolla orange to scarlet, or more usually brick-red to carmine, campanulate, somewhat oblique, 1.7-2.5 cm long, 5-lobed; tube 1-1.5 cm long, 4-7 mm in diameter at base, 8-15 mm in diameter at throat, externally more or less lepidote, internally glabrous; lobes spreading, obovate to orbicular, 6-11 mm long, 5-14 mm wide, slightly retuse or undulate at apex. Stamens 10, subequal; filaments purplish red, more or less equalling the tube, 1-1.3 cm long, glabrous; anthers brown, oblong, 2-2.5 mm long. Disk angular to lobed, glabrous. Ovary conical to ovoid, abruptly tapered to style base, about 4 mm long, densely lepidote. Style stout, 8-10 mm long, glabrous. Stigma greenish yellow, 5-lobulate within a circular rim, 2 mm in diameter. Fruiting pedicel not elongating, 1-1.5 mm in diameter. Capsule elongate-ellipsoid, 1.4-2.3 cm long, 3-7 mm in diameter, shortly beaked, inconspicuously 5-ribbed, densely lepidote with contiguous to overlapping scales. Seeds about 4 mm long including tails.

Distribution. Sumatra.

Habitat. Sub-alpine forest, above 1220 m.

Rhododendron aequabile
MARGARET STONES

RHODODENDRON PLEIANTHUM

Subgenus *Rhododendron*
Section *Vireya* subsection *Solenovireya*

A full account of this species, has been published recently in the *Kew Magazine* (Vol. 4(4), November 1987) and there is very little new information to add to our knowledge of the species. This spectacular, sweet-scented *Vireya* rhododendron is a member of subsection *Solenovireya* which, according to Sleumer's classification, comprises 33 species. Two-thirds of the subsection occur on the island of New Guinea, while the remaining third is distributed throughout Malesia. The subsection is absent from Java and the Lesser Sunda Islands. *Rhododendron pleianthum* is typical of the subsection, producing long, narrowly tubular flowers with short spreading lobes and having an indumentum of flat, scattered star-shaped scales.

This beautiful plant, for some unknown reason has been neglected by the majority of growers and ignored by authors. Nevertheless, it is a worthy addition to any collection and when grown in cultivation, is no more demanding than other *Vireya* species. Hermann Sleumer first described *R. pleianthum* in 1960 from material gathered in the Eastern Highlands of Papua New Guinea in July 1956, when Hoogland and Pullen collected it near Kerigomna Camp. One year later it was found again, this time by Robbins on Mt Hagen in the Western Highlands. Then, in the early 1960s, Schodde saw it growing in the Southern Highlands since when it has been collected from many localities on the island where, at forest margins, there are open grasslands with clumps of scrub and tree-ferns. In such localities it forms about one quarter of the vegetation. *Rhododendron pleianthum* is a terrestrial species which can eventually attain 6 m in height, but more usually reaches 4 m. It is a free-flowering plant, with large, showy pink and white flowers. These are carried in 12-15-flowered trusses measuring at least 12 cm across and are held erect on sturdy shoots. Leaves are plentiful due to comparatively short internodes, resulting in a compact shrub. Leggy growth, so much in evidence in other species of the subsection, is rarely seen in this species. Wild hybrids are known to occur between *R. pleianthum* and *R. macgregoriae*; P. van Royen collected such a plant in July 1976 in the Eastern Sepik Province of Papua New Guinea. The voucher specimen preserved at Kew has smaller leaves, and fewer, large, pink, unscented flowers than *R. pleianthum*.

Material sent to Kew by J.S. Womersley was used by Christabel King to prepare the accompanying illustration (tab. 88) for the *Kew Magazine* in June 1986. Womersley collected the plant on Mt Giluwe in the Western Highlands of Papua New Guinea.

Rhododendron pleianthum Sleumer in Reinwardtia 5: 122 (1960).

Description. Terrestrial tree or shrub 2-4(-6) m in height; branches sturdy, terete, lepidote, 5-7 mm in diameter; internodes (4-)9-13(-16) cm long. Leaves persistent, coriaceous, (4-)5(-7) in more or less patent pseudowhorls towards the top of each branch, obovate or elliptic, (5-)6.5-11(-16.5) cm long, (4-)5-10 cm wide, base cordate to rounded, apex obtuse and rounded, blade with a few dried/shrivelled scales on upper surface at maturity or scales absent, lower surface laxly lepidote, scales stellate, impressed, equal in size, more numerous near leaf margin; venation distinct, midrib flat or slightly impressed above, prominent beneath, about 3 mm wide at base, secondary veins in 8 or 9 pairs, anastomosing midway between midrib and margin. Petioles stout, grooved above, 5(-17) mm long, up to 4(-5) mm wide. Inflorescence umbellate, 10-19-flowered. Outer floral bud-scales ovate-obovate, 1.5-3 cm long; inner bud-scales oblong-spathulate, obtuse, both surfaces pubescent with dense, external median band of short, stiff or sericeous hairs; bracteoles linear, wider and more or less spathulate at tip, about 2 cm long, up to 2 mm wide, dorsally densely pubescent. Pedicels slender, 1.5-2(-2.5) cm long, densely lepidote and often short-hairy. Calyx oblique, undulate or shallowly 5-lobed, glabrous with occasional 5-7 mm long (or even longer) teeth or laciniae. Corolla pink or white, or pink and white, narrowly long-tubular, fragrant; lobes more or less patent, obovate, about 2.2 cm long, 1.2-2 cm wide, usually externally glabrous; tube 5-7 cm long, slightly arcuate and widening upwards to 10-12 mm in diameter at lobes, internally shortly pubescent, externally laxly lepidote with caducous scales. Stamens 10, unequal, slightly exceeding corolla-tube; filaments slender, hairy (sometimes glabrous at apex); anthers oblong, 4-5 mm long. Disk of ovary 5-lobed, glabrous. Ovary tapering into style, 11 × 2 mm at flowering stage, densely hairy and lepidote, the scales concealed by the hairs. Style slender, about as long as the corolla-tube, lepidote and more or less densely hairy in the basal half, laxly hairy to glabrous at apex. Stigma capitate. Capsule orange-brown, fusiform, 6-7 cm long, lepidote and hairy, valvate (usually), the valves splitting the husk as they expand, or alternatively, shedding the outer covering as an entire 'shell'; valves undulate, glossy within, externally scurfy. Seeds fusiform, about 6 mm long with a long tail at each end.

Distribution. Papua New Guinea (Western, Southern and Eastern Highlands).

Habitat. Cloud-forest and rain-forest, open grassland along forest margins, subalpine scrub on exposed ridges and mountain tops, 2600-3260 m (most frequently at about 3000 m).

Rhododendron pleianthum
CHRISTABEL KING

RHODODENDRON GOODENOUGHII

Subgenus *Rhododendron*
Section *Vireya* subsection *Solenovireya*

This species, with its narrow, long-tubular, small-lobed corollas is easily recognizable as belonging to subsection *Solenovireya*, which is distinguished from other subsections by the characteristic corolla shape. Subsection *Solenovireya* boasts about 34 species, the majority of which (about 21) were described from plants found in New Guinea. *Rhododendron goodenoughii* is included in this total and occurs on Goodenough Is., one of the volcanic group known as the D'Entrecasteaux Islands, located off the south-eastern tip of mainland Papua New Guinea.

The species was first collected by William Armit in 1895, some distance below the summit of Mt Vineuo, at 2350 m. This mountain is one of the highest points on the island and rises to 2536 m. Armit's material of the species was described by Sleumer in 1960, together with five more *Vireya* species from the Owen Stanley Range on the nearby mainland. Several other rhododendrons from the same range of mountains were described by other authors at about the same time, including one from Mt Daymann, named for William Armit (*R. armitii*). *Rhododendron armitii* is so close to *R. goodenoughii* that it is doubtful whether or not it is distinct, especially as the only characters separating the two species seem to be the prominence of the leaf-reticulation and differences in flower shape, which appear too slight to be of significance. The accompanying plate (tab. 826 new series) was painted by Margaret Stones in November 1978 from a plant raised from seed sent to Kew in 1964. It had been collected in that year by the Revd N.E.G. Cruttwell (C. 1410) who described the parent plant as a shrub 2.4 m or so tall, with waxy, pure white, *Dianthus* (Carnation) -scented flowers, growing at an altitude of about 2000 m near the top of a ridge. The species has been collected at least once since then, in August 1977, by Paul J. Kores, who found it growing in subalpine grassland at 2400 m, where it reached a height of 4 m. Given optimum conditions, *R. goodenoughii* is a strong grower, free-flowering and fragrant. The inflorescences rarely carry less than 10 flowers, which are borne erect on the sturdy branches and are admirably set off by the scattering of rust-coloured, exserted anthers and by the supporting ruff of stiff, dark green leaves. Plants are readily propagated from cuttings and can also be raised from seed.

Rhododendron goodenoughii Sleumer in Reinwardtia 5: 131 (1960).

Description. An evergreen shrub, 2.5-4 m in height with sturdy branches. Leaves in pseudowhorls of 4-8 at upper nodes, coriaceous, obovate-elliptic, 5.5-12 cm long, 3.5-7.6 cm wide; base attenuate into the narrowly winged petiole; apex obtuse, rounded or very shortly acuminate; glossy green and glabrous above, laxly scaly beneath; scales small, irregular, incised, sessile; midrib prominent on both surfaces; margin revolute. Petioles stout, flattened, grooved, up to 2 cm long, scaly. Inflorescence umbellate, 10-20-flowered. Floral bud-scales ovate, up to 3 cm long, 1.5-1.8 mm wide, caducous. Bracteoles filiform, up to 1.5 cm long. Pedicels slender, 5-12 mm long, scaly and/or minutely pubescent. Calyx small, oblique, an undulate or shallowly toothed, ciliate rim 3-4 mm in diameter. Corolla white, long-tubular with spreading lobes; tube variably curved, sometimes almost straight, narrow, scarcely widening upwards and somewhat 5-angular, 4.5-6.8 cm long, 4 mm in diameter at base, widening to 6 mm at oblique mouth of tube, externally sparsely scaly and minutely pubescent towards apex, otherwise glabrous, densely minutely pubescent within; lobes oblong-obovate, 1.3-1.5(-2) cm long, 1-1.5 cm wide, glabrous except for a few scales at the extreme base. Stamens 10, exserted from tube, but not exceeding the lobes; filaments 6.5-7 cm long, densely pubescent for at least the lower two thirds. Disk shallowly lobed and pubescent. Ovary cylindrical, gradually tapering upwards, scaly and densely adpressed yellowish-hairy, the scales concealed by the hairs. Style as wide as ovary at base, exserted and exceeding stamens, 5-6.5 cm long; scaly in lower two thirds. Stigma capitate, more or less lobed. Capsule fusiform, slightly curved, 5-6.5 cm long, with or without a beak 1 cm long, the yellowish hairs on the ovary persistent on the fruit.

Distribution. Papua New Guinea (Goodenough Is).

Habitat. Open and/or exposed situations, on or among rocks, or in grassland at high altitudes.

Rhododendron goodenoughii
MARGARET STONES

RHODODENDRON ORBICULATUM

Subgenus *Rhododendron*
Section *Vireya* subsection *Solenovireya*

This member of section *Vireya* is sometimes confused with the hardy Chinese species, *R. orbiculare* of subgenus *Hymenanthes* subsection *Fortunea*, due only to the similarity of the specific epithets. The two species are quite distinct morphologically and geographically, *Rhododendron orbiculatum* being a tropical species native and endemic to Borneo. It is a beautiful plant which produces plentiful large, pale pink, scented blooms spasmodically throughout the year.

In Sleumer's classification (1966), the subsection *Solenovireya* comprises about 33 species. These are distinguished from other subsections by the long, narrow corolla-tube, which is usually three times or more as long as the lobes, although in the case of *R. orbiculatum*, the tube often tends to be shorter than average. J.C. Moulton was the first person to collect this species, on 28 May 1911. Ostensibly the expedition was to Mt Batu Lawi (approx. 1710 m), Sabah, but according to his own account Moulton's party climbed Mt Selinguid (1455 m) on that day, starting from a camp at 930 m, which is a much more likely altitudinal range for the species than the higher elevation of Batu Lawi. This view is supported by the following note quoted by D.R. Hunt in his text to tab. 575 (new series): 'plants so labelled were not necessarily taken on Batu Lawi, but in some cases on Mt Selinguid or on the journey between the two mountains which are divided by a narrow valley'.

The plant figured here was collected in 1965 by C.J. Giles (no. 865) in Sabah, where it was growing in mossy forest at about 1200 m. It was forwarded to Kew, where, in the Himalayan House, it flowered for the first time in February 1969 and was painted by Margaret Stones for *Curtis's Botanical Magazine*.

Since 1963 Sleumer had treated *R. suaveolens* as being conspecific with *R. orbiculatum*. It was not until 1970 that D.R. Hunt, with whom other writers concurred, pointed out incontrovertible differences between the two species, both of which are now accepted at specific rank. The species most closely related to *R. orbiculatum* is *R. pneumonanthum* which, however, has elliptic or oblong leaves and more flowers in the umbel.

Rhododendron orbiculatum Ridley in J. Straits Branch Roy. Asiat. Soc. no. 63: 60 (1912).

Description. A terrestrial or epiphytic shrub or small tree up to 4 m in height. Leaves coriaceous, smooth, in pseudowhorls of 3-6, plus a few much smaller leaves, subsessile or at most shortly petiolate, orbicular or broadly ovate, sometimes elliptic, (2.4-)3-5.5(-6.6) cm long, (1.9-)4.5(-6) cm wide, with rounded to cordate base and acute, obtuse or emarginate apex, entire with a narrow, cartilaginous margin; glabrescent above, sparsely scaly beneath; scales very small, caducous, especially from upper surface. Petioles green, grooved above, 2 mm long, scaly. Inflorescence a loose umbel of (1-)3-6(-9) pale pink or cream, nodding or horizontally held fragrant flowers. Pedicels red, (1-)1.2-1.9 cm long, lengthening to 5 cm or somewhat less by fruiting stage, densely short-hairy and sparsely scaly near apex only. Calyx obsolete, an undulate or shallowly 5-lobed rim about 2 mm in diameter, laxly scaly externally. Corolla salverform; tube deep pink, straight or slightly curved, cylindrical with 5 basal pouches, only slightly widening upwards, 3.5-5 cm long, 5-7 mm in maximum diameter, laxly scaly or glabrous externally, laxly hairy in lower half within; lobes pale pink, undulate, spreading, broadly oblong or broadly obovate to suborbicular, unequal, the uppermost widest, up to 3.5(-4) cm long, 2.5-3.3 cm wide, the others up to 2 cm wide, sparsely scaly externally at base, glabrous within. Stamens 10, somewhat exserted from throat; filaments pink, slender, 4-4.5(-6.5) cm long, shortly hairy in the lower half. Disk lobed, minutely hairy on upper surface. Ovary purplish red, cylindrical below, tapering upwards into style, 5-ribbed, densely short white-hairy and with silvery scales. Style pink, exceeding stamens, about 3.5-5.6 cm long, expanded directly beneath stigma, shortly hairy and very sparsely scaly or not in lower half, otherwise glabrous. Stigma deep pink, oblique, capitate, 5-lobed, about 5 mm in diameter. Capsule narrowly cylindrical or narrowly fusiform, tapering more or less abruptly at each end, 5-ribbed, the style-base persisting as a short beak, 3-4 cm long, hairy and laxly scaly. Seeds 4 mm long (excluding tails); longest tail 1.6 mm, crimped.

Distribution. Borneo (north Sarawak, Sabah and Brunei).

Habitat. Sleumer (1966) gives the following notes: 'On Mt Kinabalu often epiphytic and with larger leaves in ridge forest between 120 and 2135 m. In N. Sarawak and Brunei rarely epiphytic, generally terrestrial and found on extreme hill "kerangas", i.e. low scrubby vegetation on bare sandstone rocks, eroding into white sand, or in elfin woodland on rugged hill crest'. Argent has given the habitat as 'an epiphyte on large trees in mossy forest but also on the ground in suitably open vegetation on mountain ridges and rocky outcrops'. Altitudinal range is (300-)800-1800 m.

Rhododendron orbiculatum
MARGARET STONES

RHODODENDRON DISTERIGMOIDES

Subgenus *Rhododendron*

Section *Vireya* subsection *Vireya* series *Linnaeoidea*

Rhododendron disterigmoides is one of the small, red-flowered, heather-like species in subsection *Vireya* series *Linnaeoidea*. So far,it has generated little interest among growers and although listed in the *Rhododendron Handbook* (1980) as being in cultivation, it is not yet available commercially. Regrettably the species is not showy enough to invoke instant acclaim, so the demand for plants is minimal. Perhaps those with no preconceived ideas of what constitutes a rhododendron might possibly see it as a 'little gem', but it would offer small reward to growers of the more spectacular examples of the genus. *Rhododendron disterigmoides* is, nevertheless, an attractive if not a spectacular plant, floriferous, with comparatively large flowers for its size and small, neat leaves. The inflorescences most frequently consist of a cluster of three flowers at the shoot-tips, while the abundance of red corollas easily makes up for their individual lack of flamboyance.

Many red-flowered plants in the mountains of Papuasia (New Guinea, Bismarck Archipelago, Solomon Islands) are thought to be pollinated by nectar-seeking birds belonging to the *Meliphagidae*, which are common in the area. It is therefore quite possible that these birds, which have been observed visiting rhododendrons in the region, are also pollinators of *R. disterigmoides*, although this species, with stamens situated all round the mouth of the corolla-tube is equally likely to attract bees. The accompanying line drawing was prepared by the author from a photograph of the holotype, Brass 9022. Brass's material was collected on Mt Wilhelm in New Guinea, at an altitude of 3225 m.

Rhododendron disterigmoides Sleumer in Reinwardtia 5: 140 (1960) and in Fl. Males. 6(4): 573, fig. 27b, c (1966).

Description. A lepidote, terrestrial shrub up to 60(-80) cm in height with robust, erect, scaly but eventually glabrescent branches; scales small, rufous, stellate, irregularly incised. Leaves coriaceous, distributed along young branches, ovate, 6.5-10 mm long, about 5 mm wide, with rounded base and obtuse, apiculate apex, margin thickened, revolute, crenulate due to the presence of impressed scales, persistently sparsely lepidote beneath, midrib and veins inconspicuous above, midrib prominent beneath. Petioles stout, 1 mm long. Inflorescence umbellate, 2-4-flowered. Pedicels slender, 8-12 mm long, densely lepidote. Calyx shortly cupular, undulate or 5-lobed, about 2.5 mm in diameter, lobes obtuse, densely lepidote. Corolla deep red, tubular, 20-23 mm long; lobes slightly flaring; tube about 1.7 cm long, 4 mm in diameter at base, widening to about 7 mm in diameter at apex, externally more or less lepidote, glabrous internally. Stamens 10, as long as, or somewhat exceeding the corolla; filaments glabrous; anthers broadly oblong, about 1.8 mm long. Disk glabrous. Ovary subconical to cylindrical, tapering gradually to style, 4-5 mm long, about 2 mm in diameter, densely lepidote. Style 8-9 mm long, glabrous. Capsule and seeds not seen.

Distribution. Western New Guinea, where it flowers in August.

Habitat. On peaty ridges, among shrubs, at about 3225 m.

Rhododendron disterigmoides
PAT HALLIDAY

RHODODENDRON SAXIFRAGOIDES

Subgenus *Rhododendron*
Section *Vireya* subsection *Vireya* series *Saxifragoidea*

Rhododendron saxifragoides belongs to the monotypic series *Saxifragoidea* of subsection *Vireya*. It is a charming plant, a native of Papua New Guinea, and is found growing in boggy areas of alpine grassland. In such situations it forms extensive cushions or mats of tightly packed, erect glossy leaves. These are studded from August to December with quite large scarlet flowers, borne high above the foliage on long pedicels. Beautiful as it is, and although Argent reports (pers. comm.) that the species flowers well in Edinburgh (and in Australia and New Zealand) *R. saxifragoides* is not easy to accommodate in cultivation, due to its need for marshy conditions, coupled with adequate drainage, in the protection of a glasshouse. Pot culture is not easy for these plants, they have such deeply penetrating roots, and there is the additional problem of keeping the soil sweet. There is no such problem in its native habitat, where it is a very successful coloniser, being locally abundant and a conspicuous feature of open places in the montane forests, as well as the preferred grassland habitats. The plant remains attractive in fruit, when the capsules split and the valves reflex. In this condition, with the persistent style remaining erect in the centre of the fruit, the plant appears to be in full bloom again, this time with red-brown flowers.

It is to be hoped that growers will develop a method of cultivation which will make this delightful plant easier to manage and more readily available. It would certainly appeal to those who appreciate rhododendrons which are a little out of the ordinary. The accompanying line drawing was prepared by the author from herbarium material of a plant collected in Papua New Guinea by Coode, Wardle & Katik (NGF 40235) in 1969 and from transparencies supplied by Tony Hall of Kew.

Rhododendron saxifragoides J.J. Sm., Meded. Rijks-Herb. no. 25: 3 (1915).

Description. A tufted, cushion or mat-forming, much-branched shrub 5-15(-30) cm in height with a 30 cm long, unbranched stem, clothed with the remains of dead leaves and with roots penetrating deep into the soil. Branches sturdy, scaly, especially near the tips, scurfy. Leaves in tufts at the ends of the short branches, dark green above, lighter beneath, more or less glossy, erect, upper surface concave, coriaceous, lanceolate or linear-lanceolate, 20-33(-52) mm long, (3-)4-5(-7) mm wide; cuneate, tapering gradually to base; apex acute-acuminate, more or less obtusely, shortly mucronate; midrib impressed above; prominent beneath, lateral veins in 2 or 3 pairs, usually inconspicuous; scales sparse above, less so beneath, small, dark and impressed; leaf margins slightly revolute. Inflorescence a solitary, deep red, scarlet or pinkish red flower which is initially nodding, but becomes erect, borne on a wine-red, unbranched, naked, laxly scaly pedicel 4.5-9.5 cm long, which is sometimes shortly hairy also; rarely, 2-flowered inflorescences occur. Calyx oblique, an undulate, scaly rim, 3 mm in diameter, sometimes shortly lobed and often ciliate. Corolla oblique, somewhat curved; tube (10-)19-25 mm long, 4.5 mm in diameter at base, expanded to 8-10 mm at mouth, externally scaly, internally pubescent; lobes scarcely spreading, suborbicular, 8-10 mm long, 4-9 mm wide, externally scaly with glabrous margins. Stamens 10, subequal, exceeding the corolla-tube; anthers purple, oblong, 2.25 mm long, 1-1.5 mm wide; filaments slender, glabrous or the lower third laxly patent-hairy. Disk lobed, glabrous. Ovary conical, about 5 mm long, 2.25 mm in diameter, tapering into style, scaly and minutely pubescent, the hairs obscuring the scales. Style up to 15 mm long, stout, equalling or slightly shorter than the stamens, thickening in fruit. Stigma enlarging somewhat in fruit, up to 2 mm in diameter. Fruiting pedicel up to 10.5 cm long. Capsule dark wine-red, finally brown, fusiform, erect, about 2 cm long, up to 2 mm in diameter, dehiscing by splitting into valves which reflex around the persistent style and stigma, resembling open flowers, encrusted with scales and minutely pubescent. Seeds about 2.5 mm long, with a tail at each end.

Distribution. New Guinea. (Along the main mountain range, from the Oranje Mts in Irian Jaya to Mt Hagen in Papua New Guinea. The species is possibly even more widespread than these records suggest).

Habitat. Boggy areas in montane grassland and open places in montane forest, (3180-)3400(-4000) m.

Rhododendron saxifragoides
PAT HALLIDAY

RHODODENDRON TAXIFOLIUM

Subgenus *Rhododendron*
Section *Vireya* subsection *Vireya* series *Taxifolia*

In the dried state this epiphytic species is strongly reminiscent of the closely allied *R. stenophyllum*. The difference between the respective geographic distributions of these two species, together with the taxonomic discrepancies (e.g. white flowers in *R. taxifolium*, orange in *R. stenophyllum*), convinced Professor Sleumer that these species should be kept separate. He therefore created a monotypic series to accommodate *R. taxifolium*. He used the narrower leaves of *R. taxifolium* and its greater number of leaves in each pseudowhorl to further distinguish it from *R. stenophyllum*. First collected in 1925 in the Philippines by M. Ramos and G. Edaño, as yet *R. taxifolium* is not in general cultivation, although it no doubt exists in a few specialist collections. There are only two sheets of the taxon in Kew herbarium, both from Mt Pulog in Benguet Province, Luzon, to which area, according to Sleumer (1966) it is apparently confined. *Rhododendron taxifolium* is a beautiful plant, rarely exceeding 1 m in height which, in the wild, blooms in February and March when it produces several quite large white flowers with contrasting dark brown anthers, set in a ruff of dark narrow leaves at the ends of the branches. It is a pity that the plant has not become available to would-be growers for, like *R. stenophyllum*, in a variable genus it stretches variability to the uttermost, thereby acquiring an attraction all its own. The line drawing of this species was prepared by the author from a herbarium specimen collected by Ramos and Edaño in 1925 (R & E 44880).

Rhododendron taxifolium Merr. in Philipp. J. Sci. 30: 419 (1926).

Description. An epiphytic shrub up to 1 m in height with slender, glabrous or only sparsely scaly branches; internodes about (1-)3-4 cm long at base of stems, but much shorter near the top of the plant. Leaves coriaceous, stiff, 20 or more in dense, congested pseudowhorls at upper nodes, more or less glossy on both surfaces, linear, very narrow, flat or revolute, 2.2-4.2(-4.8) cm long, 1.5-2 mm wide, base cuneate, tapering into a short ill-defined petiole, or sessile, apex obtuse to acute, margin crenulate, midrib impressed above, prominent beneath, upper surface very sparsely scaly or glabrous, lower surface papillate with more numerous scales; scales small, impressed. Flowers scentless, solitary or in clusters of 2 or 3 at the ends of the branches, closely surrounded by the terminal tufts of leaves. Pedicels slender, about 1 cm long, hairy and laxly scaly. Calyx a scaly and minutely pubescent rim or shortly 5-lobed, 4 mm in diameter. Corolla white, tubular-campanulate, 2-2.1 cm long to lobe-tip, 2-2.5 cm in diameter (in dried material); tube externally scaly, internally pubescent; lobes 5, ovate, 10 mm long, 8 mm wide, with very few scattered scales. Stamens 10, subequal, equal or shorter than the corolla-tube; filaments linear, pubescent for most of their length. Disk glabrous. Ovary about 3 mm long, densely white-hairy and sparsely scaly. Style 5 mm long, hairy near base. Capsule about 1 cm long.

Distribution. Philippine Islands (Luzon, on Mt Pulog).

Habitat. Sandy-loamy soil in primary forest, 2300-2700 m.

Rhododendron taxifolium
PAT HALLIDAY

RHODODENDRON STENOPHYLLUM

Subgenus *Rhododendron*
Section *Vireya* subsection *Vireya* series *Stenophylla*

Not many rhododendrons bear such narrow leaves as this one and at first glance it is not easy to recognize *R. stenophyllum* as belonging to the genus. Argent (1988) included it in his series *Vireya*, but it is here accepted as a member of series *Stenophylla* Sleumer (1960). It is related to *R. nervulosum* and *R. exuberans* in the same series, both of which bear flowers similar in shape to those of *R. stenophyllum*, but differ in their broader leaves. Although collected by Sir Hugh Low in Sabah as early as 1867, and named by J.D. Hooker at that time, the name *R. stenophyllum* was not validated until Stapf published a paper in 1894 on the flora of Mt Kinabalu and cited Low's specimen. F.W. Burbidge, one of Veitch's plant collectors, may have been the first person to introduce live material of this plant, although no record of the introduction exists. He had mentioned and illustrated the species previously in his book *The Gardens of the Sun*, which was published in 1880, but the plant had not attracted much attention. Even today, it is rare in cultivation. One would expect that the unusual appearance of *R. stenophyllum* would recommend it to growers; apparently it does not! Perhaps this is because the species so rarely produces a heavy crop of flowers, although subsp. *stenophyllum* is more satisfying in that respect than subsp. *angustifolium*. Whatever the reason for the gardeners' apparent lack of enthusiasm, *R. stenophyllum* has certainly stimulated the interest of artists, photographers and collectors in more recent years. In 1961, Mrs S. Collenette collected and photographed it in Sabah; it was then brought back from Kinabalu by H. Sleumer in 1963, and again by W.R. Price in 1967, who also photographed it. It is to be hoped that, being easy to propagate, growers will soon recognize its worth as a greenhouse plant and produce enough plants to make the species readily available commercially. J.J. Smith first separated the exaggeratedly narrow-leaved variant of *R. stenophyllum* in 1935 and named it var. *angustifolium*. Argent, Lamb and Phillipps (1984) raised the variety to subspecific rank.

Subsp. *stenophyllum* is distinguished from subsp. *angustifolium* by its broader leaves and scaly, sparsely hairy, usually shorter pedicels. Subsp. *angustifolium* has narrow leaves 30 times (or more) longer than wide and softly hairy pedicels 8-18 mm long. It also grows at lower altitudes than subsp. *stenophyllum*. The plants collected by Sir Hugh Low and Burbidge were both referred to subsp. *stenophyllum*.

The present plate was painted by Rodella Purves in Edinburgh from material cultivated at the Royal Botanic Garden there. It was published in the *Kew Magazine* (tab. 105).

Rhododendron stenophyllum Hook. fil. ex Stapf in Trans. Linn. Soc. London, Bot. 4: 196 (1894) subsp. **stenophyllum**.
R. stenophyllum subsp. **angustifolium** (J.J. Sm.) Argent, Lamb & Phillipps in Notes Roy. Bot. Gard. Edinburgh 42: 115 (1984).
R. stenophyllum var. *angustifolium* J.J. Sm. in Bull. Jard. Bot. Buitenzorg ser. III, 13: 452 (1935).

Description. Usually terrestrial, sometimes epiphytic, sparsely branched shrub, 0.6-1(-3) m in height, with a slim trunk and green, smooth, slender, terete branches; young growth sparsely scaly, internodes 2-10 cm long. Leaves subsessile, in pseudowhorls of (8-)10-15, persistent at uppermost 2-5 nodes, coriaceous, glossy, dark to medium green, linear, narrowly linear or subsetaceous, (4-)5-7.5(-12) cm long, (1-)3(-6) mm wide, base attentuate, apex more or less acute to subpungent, margin entire, not revolute, upper surface sparsely scaly, finally glabrescent, mid-green, lower surface sparsely scaly, midrib impressed above, prominent beneath, secondary venation of about 7 pairs of inconspicuous lateral veins; scales small, flat, suborbicular with dark centre, not impressed. Petioles indistinct, green, short, stout, grooved above, 2(-4) mm long, sparsely scaly. Inflorescence umbellate, with 1-3 nodding or horizontally held flowers. Floral bud-scales reddish brown, long-attenuate, somewhat spreading, up to 2.8 cm long, 8 mm wide. Pedicels red, stout, 6-18 mm long, 1-1.5 mm in diameter, laxly scaly and more or less densely, softly pubescent. Calyx fleshy, an entire, undulate disk or shortly, bluntly 5-lobed, about 4 mm in diameter, laxly lepidote, pubescent at base. Corolla waxy, bright to dark red, or clear orange-scarlet, or orange with red base, narrowly campanulate to broadly funnel- or bell-shaped, 2-3.2 cm long, about 4.5 cm across lobes, glabrous throughout; tube 22 mm long, 4-6 mm in diameter at base, (8-)13-16 mm at throat (= base of lobes); lobes broadly obovate-spathulate, more or less spreading, about 12 mm long, 10 mm wide. Stamens 10, arranged in a ring around the throat of the corolla and subequalling the corolla-tube; filaments orange or red, linear, 12-15 mm long, sparsely pubescent at base; anthers cream, 3-5 mm long. Disk conspicuous, glabrous. Ovary cylindrical to subovoid, 4-5 mm long, 2.5 mm in diameter, densely shortly white-hairy and minutely lepidote. Style abruptly tapering to ovary, 6.5 mm long, elongating to 13 mm at fruiting stage, glabrous. Stigma cream, somewhat lobed, 2.5 mm in diameter. Capsule reddish purple, more or less cylindrical, slightly arcuate, about 18 mm long, 5 mm in diameter, sparsely, minutely lepidote and densely short-hairy. Flowering is spasmodic throughout the year, but mainly occurs from February to June in subsp. *stenophyllum*, and September to April in subsp. *angustifolium*.

Distribution. Subsp. *stenophyllum*: Sabah, restricted to Gunong Kinabalu, 2700-2800 m. Subsp. *angustifolium*: Borneo, Sabah, Sarawak, Brunei and Kalimantan. On Gunong Kinabalu it is found at lower altitudes than the type subspecies (1500-2400 m).

Habitat. Both subspecies grow in mossy submontane forest in fairly open situations.

Rhododendron stenophyllum
RODELLA PURVES

RHODODENDRON CITRINUM var. CITRINUM

Subgenus *Rhododendron*
Section *Vireya* subsection *Vireya* series *Citrina*

Botanical Magazine tab. 4797 represents one of the few subtropical rhododendrons painted by W.H. Fitch. Hasskarl named and described *R. citrinum* in 1844 and reported that it was growing epiphytically on the trunks of old trees in the mountains of Tjiburrum, at 1600 m (approx.) above sea-level. Henshall, who was collecting plants in Java for Messrs Rollisson of Tooting, 10 years later, extended the range to 3200 m, but was unable to find the plant above that altitude.

The species belongs to the monotypic series *Citrina* of subsection *Vireya*. Sleumer (1966) recognized two varieties; var. *citrinum* from west Java and Bali, with yellow or yellowish flowers, and var. *discoloratum* from Sumatra, with, usually, orange or scarlet (rarely yellow) flowers. Tab. 4797 shows var. *citrinum* and was painted from plants collected by Henshall and supplied by Messrs Rollisson in May 1854. The pale, fragrant flowers contrast strongly with the dark green, glossy foliage; they are small, creamy or lemon-yellow, borne on red pedicels and are accentuated by the 5 (never 10) orange stamens — definitely not a showy species, but a *Vireya* with a somewhat more modest charm than many of its allies. In the wild the flowers are produced from September to March. In cultivation *R. citrinum* responds to the same treatment and conditions as are provided for others of its subsection.

Rhododendron citrinum (Hassk.) Hassk., Cat. hort. bot. Bogor 161 (1844) var. **citrinum**.
Azalea citrina Hassk., Flora, Beibl. 25: 30 (1842).
Rhododendron zippelii Blume, Fl. Jav. Pl. (*ined.*) t. 4 (1863-83).
R. jasminiflorum Koord. Junghuhn Gedenkb. 184 (1910), *non* Hook. (1850).

Description. A small, evergreen shrub up to 2 m in height in the wild, with slender, scaly, rust-coloured branches. Internodes 2.5 cm long. Leaves in pseudowhorls of 4 or 5 at the upper 1-2(-3) nodes, coriaceous, unequal, dark, glossy green above, greenish with brown scales beneath, elliptic, oblong-elliptic to lanceolate, (1.5-)2.5-5(-6.5) cm long, (1-)1.5-2.7(-4) cm wide, with attenuate base and broadly acuminate or obtuse apex. Petioles 4-9(-12) mm long, scaly. Inflorescence a (1-)2-4(-5)-flowered, terminal umbel. Pedicels red, slender, up to 1.5 cm long, puberulent and scaly. Calyx 3-4 mm in diameter, oblique, with 5 obtuse, externally scaly lobes. Corolla nodding, more or less fragrant, campanulate with 5 rounded lobes, lemon, pale or creamy yellow, 1.5-1.7 cm long, externally more or less scaly (more densely so in Balinese plants). Stamens 5, unequal, as long as, or slightly exceeding the corolla-tube; filaments red or orange, linear, glabrous; anthers orange. Ovary oblique, ellipsoid, approximately 4 mm long, rather abruptly tapering into the style, glabrous, with undulate, glabrous, basal disk. Style slender, expanding upwards, 5-6 mm long, approximately equalling the stamens in length. Stigma convex, lobulate. Capsule more or less fusiform, slightly curved, 2-2.5 cm long.

Distribution. Var. *citrinum* grows in Indonesia (western Java, Bali).

Habitat. Very humid, subalpine primary forests or on their margins, (1000-)1450-2400(-3200) m. Var. *discoloratum* occurs in Sumatra, in mossy forests on ridges, between altitudes of 1220 and 2500 m.

Rhododendron citrinum var. *citrinum*
WALTER HOOD FITCH

RHODODENDRON BURTTII

Subgenus *Rhododendron*
Section *Vireya* subsection *Vireya* series *Buxifolia*

This delicate species is named after Mr B.L. Burtt who has made extensive plant-collections and who for many years worked in the herbarium at Kew before going north to the Royal Botanic Garden at Edinburgh. The series *Buxifolia* is a scattered one, with representatives in Sumatra, the Philippines, New Guinea, Celebes, the Malay Peninsula and Borneo. *Rhododendron burttii* has been found in both Sarawak and Sabah in Borneo. The material from which the original description was drawn up was collected by Mr Burtt in Sarawak in 1967, on the route from Bakelalan to Gunong Murud. It was grown on at the Royal Botanic Garden, Edinburgh, where herbarium specimens were later deposited. Since then it has been collected from Mt Kinabalu by John H. Beaman et al. (January 1984) at an altitude of between 1300 and 1600 m.

A plant grown by Dr Charles Nelson of the National Botanic Garden, Glasnevin, in his private collection, was the subject of tab.130. It was painted by Wendy Walsh for the *Kew Magazine*.

The species most closely allied to *R. burttii* appears to be *R. frey-wisslingii*, although on examining the two species, the relationship is not immediately obvious. *Rhododendron frey-wisslingii* bears dense tufts of narrowly oblanceolate, obtuse to rounded leaves at the tops of the branches, and is a twiggy plant with very short internodes. *Rhododendron burttii* by contrast, is straggly, with longer internodes and the leaves, therefore, are more distinctly whorled. The flowers of the two species are dissimilar, those of *R. frey-wisslingii* are broadly and shortly tubular, glabrous with short, apparently erect, lobes, while the corollas of *R. burttii* are campanulate with externally hairy, spreading lobes. These are the main differences; there are other lesser but supportive characters which serve to strengthen the relationship between the two species.

Rhododendron burttii P.J.B. Woods in Notes Roy. Bot. Gard. Edinburgh 37: 157 (1978).

Description. Straggly terrestrial or epiphytic shrub up to 0.5 m, erect or scrambling, with slender, often decumbent branches, about 3 mm in diameter, internodes 3-6(-10) cm long; young growth pubescent and laxly scaly. Leaves coriaceous, dark green above, more or less glossy, obovate-oblanceolate or elliptic, 25-30(-38) mm long, (9-)11-13 mm wide, base cuneate, apex obtuse, margin hardly recurved, midrib impressed above, prominent beneath, scales with broad, erose margins, not impressed, except on lower leaf surface. Petioles more or less distinct, (2-)3(-4) mm long, pubescent and sparsely scaly. Inflorescence of 1-2(-5), nodding or horizontal flowers. Pedicels purplish red, (12-)20-25 mm long, lengthening in fruit to about 27 mm, 1 mm in diameter, with long white hairs. Calyx an undulate or obscurely 5-lobed rim, ciliate, long-hairy at base only. Corolla dark red, reddish purple or bright scarlet, tubular-campanulate, 28-31 mm long (including lobes), 20-22 mm in diameter (i.e. across lobes from tip to tip), tube 17-20 mm long, 6-8 mm in diameter at throat, base of tube pubescent internally, lobes spreading, somewhat reflexed, oblong, obtuse, 9-11 mm long, 3.5-6 mm wide, with long white hairs externally and with some similar hairs at base of lobes on inner surface. Stamens 10, subequal, usually shortly exserted and arranged in a circle in the mouth of the corolla-tube; filaments purplish red, linear, dilated and pubescent at base, 18-20 mm long; anthers orange-yellow, oblong, about 1-1.5 mm long. Disk glabrous. Ovary 3-4 mm long, densely white-hairy, also scaly, the hairs concealing the scales. Style slender, straight, 10-11 mm long, long-hairy in basal two thirds and sparsely scaly. Stigma 1-1.75 mm in diameter. Capsule (immature) cylindrical, probably finally fusiform, (11-)13-17 mm long (excluding persistent style), 3-6 mm in diameter, densely long-hairy and laxly scaly. Seeds not seen.

Distribution. Borneo (Sarawak and Sabah).

Habitat. Montane forest where it occurs as an undershrub, about 1300-1600m.

Rhododendron burttii
WENDY WALSH

RHODODENDRON WRIGHTIANUM

Subgenus *Rhododendron*
Section *Vireya* subsection *Vireya* series *Buxifolia*

Rhododendron wrightianum was introduced into cultivation through the agency of Dr H. Sleumer and the Rijksherbarium, Leiden. It had been discovered in 1909 and formally described in 1912 by S.H. Koorders, who named it for the then Assistant Keeper of Kew herbarium, Charles Henry Wright (1864-1941).

The species is a native of the mountains of western New Guinea and also occurs in the south-eastern corner and offshore islands of that country. In the wild it flowers from June to October. Dr Sleumer collected the species on Mt Cycloop at 1640 m and the accompanying illustration was painted for *Curtis's Botanical Magazine* by Margaret Stones in March 1973, from young plants received at Kew in 1967. These plants derived from Sleumer's collection of var. *cyclopense* and were propagated at the Strybing Arboretum in San Francisco. Several varieties of the species have been described, but only one of them (var. *insulare*) seems to be distinct enough to merit recognition, and this is the form found on Normanby and Goodenough Islands, both located off the south-east coast of New Guinea. At the time tab. 653 (new series) was painted, var. *cyclopense* was considered to be distinct on the basis of the indumentum of the pedicels and discrepancies in leaf-size. These differences have since proved to be unreliable and the variety has accordingly been reduced to synonymy under *R. wrightianum*.

Rhododendron wrightianum is one of the smaller *Vireya* rhododendrons, with neat leaves and waxy, translucent, rich red flowers. The small, box-like leaves place it in the series *Buxifolia* which also includes *R. burttii*, another small-leaved, red-flowered species (see plate 15, above).

It is interesting to note that of the 41 species included in the series *Buxifolia* by Sleumer (1966), 27 are endemic to New Guinea.

Rhododendron wrightianum Koord., Nova Guinea 8(2): 880 (1912).
R. wrightianum var. *cyclopense* J.J. Sm., Nova Guinea 12: 130 (1914).
R. wrightianum var. *ovalifolium* J.J. Sm., *loc. cit.* 131, t. 29B.

Description . A small, diffuse shrub up to 1 m tall, with long, slender branches, terrestrial or epiphytic on both dead and living trees; older bark peeling and minutely verrucose. Leaves in pseudowhorls of 3-6 at upper nodes, separated by internodes (1-)2.5-6 cm long, coriaceous, glossy, obovate-oblanceolate, sometimes almost orbicular, (1-)23-30(-42) mm long, 10-15(-20) mm wide, tapering to cuneate base, apex obtuse to rounded, retuse or sometimes with minute mucro, margin entire, slightly recurved, midrib prominent beneath, other nervation inconspicuous, blade more or less densely scaly above (where the scales are 3-4 times their own diameter apart), smooth and paler in colour beneath and laxly scaly; scales minute, darker than the leaf-surface, impressed. Petioles 1-3 mm long. Inflorescence terminal, of 1-2(-3) nodding flowers; Floral bud-scales persistent, broadly ovate, apiculate, 4-8 mm long, minutely silky, ciliate with stellate, dendroid scales. Bracteoles filiform or with flattened apex, up to 8 mm long. Pedicels 12-15(-23) mm long, lengthening in fruit, slender, scaly, scales stellate, dendroid. Calyx cupular, oblique, about 3 mm in diameter, undulate to irregularly 5-lobed, densely scaly at base with stellate, dendroid hairs. Corolla waxy, translucent, pink, red or crimson but usually deep ruby red, tubular with spreading lobes, approximately 16 mm across lobes; tube slightly curved, widening upwards, 14-20(-23) mm long, 4 mm in diameter at base, 7-8 mm in diameter at throat; lobes 6-7(-10) mm long; tube and centre of lobes externally laxly stellate-scaly, glabrous within. Stamens 10, subequal, as long as corolla-tube or slightly exserted from it; filaments linear, glabrous; anthers approximately 1.3 mm long. Disk distinctly (?10-) lobed, glabrous. Ovary cylindrical to fusiform, tapering into style, 3-4 mm long, 1 mm wide, densely scaly. Style slender, (8-)13-15 mm long (8-10 mm long in material figured), perhaps lengthening in fruit, glabrous except for extreme base which bears persistent scales. Stigma oblique, clavate-capitate, at least 1 mm in diameter. Mature capsule not seen, young fruit more or less fusiform, beaked, 18 mm long, 2 mm in diameter.

Distribution. East and west New Guinea and its offshore islands. The variant figured here is confined to Mt Cycloop.

Habitat. Alpine moss-forest, or on crests of mountain ridges in more or less open situations, 1480-2160 m.

Rhododendron wrightianum
MARGARET STONES

RHODODENDRON MACGREGORIAE

Subgenus *Rhododendron*
Section *Vireya* subsection *Vireya* series *Javanica*

Although many *Vireya* rhododendrons have become commonly cultivated only in the last two decades or so, they are not new to horticulture. Indeed, some species were grown by enthusiasts and hybridists as long ago as the late 1800s but their popularity did not last and the group was largely ignored until recently. Nowadays, many exciting species, the products of numerous collecting expeditions to South-east Asia, are becoming available from specialist nurseries.

The series *Javanica* is distinguished from the other series within the subsection *Vireya* (syn. *Euvireya*) by its comparatively large, broad leaves, borne on erect shrubs or small trees. Although discovered nearly 100 years ago, *R. macgregoriae* is a quite recent introduction into cultivation. It was distributed as seed, from the Rijksherbarium, Leiden, sometime after 1955, but the circumstances and source of the collection are not known. The species was first collected in New Guinea on Mount Yule at an altitude of 3350 m in 1890/91, by a Mr Belford, whose dried material was described by Baron von Mueller. The plant was named for Lady Macgregor, wife of Sir William Macgregor, the first Lieutenant-Governer of New Guinea.

The material from which the accompanying plate (tab. 552 new series) was painted by Margaret Stones in February 1967, was collected by Mr C.D. Sayers along the Okapa road, 24 km west of Kainantu in Papua New Guinea, at an elevation of 1550 m. This plant was growing in open grassland, but the species is widespread in New Guinea and also occurs in primary mossy forest, or among trees in dry or moist, sunny or lightly shaded situations.

Rhododendron macgregoriae is known to have poisonous properties which are utilized by the local population; rat-bait is prepared by mixing the flowers with sweet potato. It is a common and extremely variable species, especially in flower colour, which can be light yellow, orange or red, with or without a central 'eye' of yellow or orange, and flowers are produced throughout the year. It is known to produce natural hybrids in its native habitat, either *R. zoelleri* (another common species), *R. dielsianum* or *R. inconspicuum* frequently being the other parent. The varieties recognized by Sleumer are based on the presence or absence of indumentum on the corolla and filaments, but whether or not this character can be considered a firm distinction between the varieties is debatable.

Rhododendron macgregoriae is thought to be pollinated by butterflies, which were observed visiting the flowers by M. Black. Professor Peter F. Stevens noted a species of *Papilio* similarly occupied.

Like other epiphytic rhododendrons, *R. macgregoriae* requires a very open compost and it is best grown in a raised bed to facilitate free drainage. Night temperatures should be maintained at a minimum of 7-13°C, and although the atmosphere should be kept humid, it should allow for free movement of air around the plant, to minimize the risk of mildew. Seeds germinate readily, and if not sown immediately, can be kept in cold storage for about one year. Cuttings of semi-ripe material can be rooted under mist with bottom heat.

Rhododendron macgregoriae F. v. Mueller in J. Bot. 29: 177 (1891).
R. hansemannii Warb. in Bot. Jahrb. Syst. 16: 26 (1892).
R. vonroemeri Koord. in Nova Guinea 8: 879, t. 155 (1912).
R. calceolarioides Wernh. in Trans. Linn. Soc. London, Bot. 9: 95 (1916).
R. gorumense Schlechter in Bot. Jahrb. Syst. 55: 142 (1917).

Description. A terrestrial (rarely epiphytic) shrub or tree up to 5 m (rarely to 15 m), with slender branchlets in whorls of (2-)3(-5), scaly, finally glabrous; internodes (2-)4-10(-15) cm long. Leaves coriaceous, very shortly petiolate, in pseudowhorls of 3-5(-8) at upper nodes, glossy dark green above, paler beneath, ovate-lanceolate or elliptic-oblong, (3-)4-8(-17) cm long, (1.5-)3.5(-6) cm wide, broadly attenuate to base, rarely rounded, apex acuminate to obtuse, finally more or less elepidote above, densely scaly beneath, scales irregularly incised, midrib impressed, lateral veins not impressed above, in 6-10 pairs, reticulations often evident. Petioles (3-)6-12 mm long, scaly. Inflorescence a dense umbel of (4-)8-15 flowers. Pedicels pink to red, usually slender and sparsely scaly, occasionally laxly pubescent, (1.5-)2.5-4(-7) cm long. Calyx oblique, undulate, ciliate. Corolla light yellow to dark orange, often reddish in throat; tube straight, about 1 cm long, with 5 nectar-pouches at base, 3-5 mm in diameter throughout, laxly scaly externally, hairy at base within; lobes spreading, obovate-spathulate, 1.2-1.8(-2) cm long, 0.8-1.3(-1.6) cm wide, laxly scaly externally. Stamens 10, unequal, subequalling the corolla; filaments subvillous in lower half, tapering upwards; anthers 2.5-3.2 mm long. Disk lobed, usually glabrous, occasionally pubescent on upper surface. Ovary long-conical, tapering into the style, about 4 mm long, densely adpressed-hairy and scaly, the hairs concealing the scales. Style persistent, 1-1.5 cm long, more or less hairy and scaly in lower half, glabrous above. Stigma capitate with distinct lobes. Capsule said to be dark orange, cylindrical to fusiform, somewhat arcuate, 5-ribbed, 3.5-4.5 cm long, 3-4 mm wide, with contorted valves, hairy and scaly. Seeds about 4 mm long, including the tail at each end.

Distribution. Papua New Guinea.

Habitat. Open situations, moss forest and grassland, (120-)500-3000(-3350) m.

Rhododendron macgregoriae
MARGARET STONES

RHODODENDRON ZOELLERI

Subgenus *Rhododendron*
Section *Vireya* subsection *Vireya* series *Javanica*

Of all the *Vireya* rhododendrons, *R. zoelleri* must surely be the most flamboyant, with red, orange and yellow predominating in its flowers which are produced spasmodically all the year round. Not only are the colours of the flowers unsurpassed in brilliance, but the size of the corollas is, in the best forms, equally remarkable. These are thought to be pollinated by butterflies (*Ornithoptera* sp., *Papilio* sp.) which were observed visiting the flowers in the daytime by Docters van Leeuwen (1926). The species is widespread and locally abundant and hybrids with *R. macgregoriae*, another widespread species, are known to occur. Sleumer (1966) reports that *R. zoelleri* also hybridizes in at least one locality in Papua New Guinea with *R. phaeopeplum*, the suspected progeny of such crosses producing white or yellowish flowers. Sleumer was the first to discover these putative hybrids in Papua New Guinea, in an area where both parents are abundant. It is therefore not surprising that *R. zoelleri* has been used by hybridists although the offspring rarely surpass their spectacular parent. There are numerous clones of this popular species, one of which, 'Decimus', the clone depicted by Margaret Stones in tab. 682 (new series) received an Award of Merit when exhibited from the Royal Botanic Gardens, Kew, at an RHS Show in 1973. The name commemorates Decimus Burton who designed the Temperate House at Kew where the plant was grown. It was raised from seed collected by Rosendahl (BW 3251) in the Wissel Lakes region of west New Guinea at 1800 m altitude, and distributed by the Rijksherbarium, Leiden in 1955. Most of the other cultivars bear flowers which glow with brilliant, almost fluorescent colour. F.C. Hellwig, during an expedition to the Finisterre Mountains of north-east Papua New Guinea in October 1888, first collected *R. zoelleri* while he was travelling with Hugo Zöller, the leader of the expedition. Zöller was a correspondent of the Kölnische Zeitung and author of Deutsch-Neu-Guinea und meine Ersteigung des Finisterres Gebirges (1891). The plant was not named until the following year, when Otto Warburg published the name, with a formal description, in Engler's Jahrbücher. The series *Javanica* is a large group within subsection *Vireya*, comprising some 79 or 80 species and including *R. lochae*, the only native Australian Rhododendron. Closely allied to *R. zoelleri* is *R. laetum*, which differs in its broader, subsessile, coriaceous leaves, shorter anthers and yellow flowers, which, although they later become orange or reddish, do not even approximate to the intensity of colour found in the flowers of *R. zoelleri*. *Rhododendron macgregoriae* is another close relative which at first glance seems very much like a small version of *R. zoelleri*, but the leaves are in whorls of 3-5, the more numerous flowers in each umbel have very short corolla-tubes (about 1 cm long), the corolla is lobed for a half to two-thirds of its length and the anthers are considerably smaller.

Rhododendron zoelleri Warb. in Bot. Jahrb. Syst. 16: 15, 24 (1892).
R. asparagoides Wernh. in Trans. Linn. Soc. London, Bot. 9: 94 (1914).
R. moszkowskii Schlechter in Bot. Jahrb. Syst. 55: 161 (1918).
R. oranum J.J. Sm. in Feddes Repert. 30: 167 (1932).
R. doctersii J.J. Sm., Nova Guinea 18: 98, tab. 21, 1 (1936).
R. laetum J.J. Sm., Nova Guinea 18: 98 (1936) *non* J.J. Sm (1914).

Description. A lax, evergreen shrub or small, erect tree to about 6 m, occasionally more; most frequently terrestrial, sometimes epiphytic; branches of current season green and scaly, older branches brown with a reddish tinge; internodes (3-)5-10(-14) cm long. Leaves subcoriaceous in whorls of 3-6(-7) (to 12 in material figured) at tips of young branches, narrowly to broadly elliptic or lanceolate, 7-17.5 cm long, (including petiole), 3.8-8.5(-10) cm wide, rounded to attenuate at base, acute to acuminate at apex; margin entire; midrib grooved at base on upper surface, prominent beneath, main veins 10-12(-14) pairs; upper surface dark glossy green and densely scaly, lower surface pale green, densely stellate-scaly, the scales on both surfaces small, brownish and caducous, both leaf surfaces finally glabrescent and punctulate. Petioles more or less stout, grooved above, 6-13(-20) mm long, scaly. Inflorescence a fragrant or scentless umbel of 5-8 flowers. Pedicels pallid yellow-green, suffused with pink or red, comparatively slender for their length, (2-)3-4 cm long, (to 5 cm long in fruit), laxly stellate-scaly and minutely pubescent. Calyx obsolete, an oblique, undulate or 5-toothed pallid yellow-green rim 2-4(-5) mm in diameter, laxly scaly and minutely pubescent. Corolla funnel-shaped with 5 spreading lobes; 5-9.5 cm long overall and 7-9 cm across lobes; tube 2.5-3.5(-4) cm long, straight, 5-pouched and 5-8 mm in diameter at base, 10-20(-22) mm in diameter at throat, lemon to butter-yellow, sparsely white-scaly externally, pubescent at base within; lobes obovate-spathulate, 2-2.3(-5) cm long, 1.8-2(-4) cm wide, rounded or retuse, yellow at base, pinkish red, salmon or orange to darker red towards apex. Stamens 10, subequal, filaments yellow to orange, (4-)5-5.5 cm long, pubescent in basal third or half; anthers golden yellow, reddish-brown or purple, 4-5(-8) mm long. Disk more or less lobed, 1.2 mm thick, densely white-pubescent on upper surface. Ovary more or less cylindrical but tapering to style, 1-1.8 cm long, densely tomentose with short adpressed white hairs obscuring minute scales. Style green or yellow, capitellate, (4-)5-lobulate. Capsule fusiform or cylindrical, somewhat curved, hairy and scaly, (6.5-)18-19 cm long. Seeds about 8 mm long, with a long tail at each end.

Distribution. New Guinea and west Ceram in the Moluccas. It has not been found on any other islands in that group.

Habitat. A variety of habitats and soils (growing epiphytically and terrestrially), sea-level to 1500(-2000) m.

Rhododendron zoelleri
MARGARET STONES

RHODODENDRON IMPOSITUM

Subgenus *Rhododendron*
Section *Vireya* subsection *Vireya* series *Javanica*

Seeds and young plants of the handsome white-flowered variant of this red, pink or white-flowered species were collected by a Kew botanist, Martin J.S. Sands (Sands 457), in November 1969 in the Latimodjong Mountains of south-west central Sulawesi (Celebes). The plants were growing at an elevation of 2750 m near Buntu Jakke, one of the ridge-tops of that mountain chain. The accompanying plate (tab. 816 new series) was painted by Christabel King from Sands' material in February 1980. This was not the first time the species had been collected. The type, which has red flowers, was described by J.J. Smith in 1937, and was found at Ebenda, at an elevation of 2500 m, in June 1929 by G. Kjellberg (no. 4149). Kjellberg also gathered material of the same species (no. 3926), at Buntu Poka Pinjang, between 2400 m and 2700 m in the same year.

The specific epithet *impositum*, refers to the abrupt junction between style and ovary in this species. It is distinguished from other members of series *Javanica* by its relatively small, thick, broad leaves which, instead of being in clusters or pseudowhorls, are spaced along the branches, and by the glabrous or glabrescent style.

Rhododendron impositum J.J. Sm. in Bot. Jahrb. Syst. 68: 201 (1937).

Description. Terrestrial shrub or small tree straggling to 4 m, with rather stiffly held branches; young growth sparsely scaly, soon becoming glabrescent. Leaves dark green and glossy above, paler beneath, not whorled but alternate and spaced along the branches, thick, coriaceous, oblong or broadly elliptic, occasionally lanceolate and acuminate, (3.7-)5-9(-10) cm long, (2.3-)3-4(-5) cm wide, rounded at base, rounded and abruptly acute to acuminate at apex, margin revolute, midrib broad at base, prominent and raised on both surfaces, veins in 6-9 pairs, conspicuous only on lower surface; densely scaly, but scales caducous above, more persistent beneath, leaves finally glabrescent. Petioles green or stained with pink, stout, flat, grooved above, 5-8(-10) mm long, 2-3 mm in diameter, scaly at first, later glabrescent. Inflorescence a terminal umbel of 5-8(-15) flowers. Floral bud-scales obovate-spathulate, about 3 × 1.8 cm, externally silky-hairy. Bracteoles filiform, linear, 1-1.5(-2) cm long, pilose. Pedicels pink, scaly, 2-3 cm long. Calyx obsolete, an oblique, undulate rim about 4 mm in diameter with an occasional longer lobe to 5 mm long. Corolla white, pink or red, funnel-shaped, sparsely externally scaly; tube straight, 1.2-2 cm long, pouched at base, 5 mm in diameter at base, widening upwards to 15 mm in diameter at throat, internally pubescent in lower part; lobes 5, spreading, oblong-obovate, retuse, (1.2-)2.5-3 cm long, (1.3-)2 cm wide. Stamens 10, subequal, as long as the tube or slightly exserted; filaments 2.5 cm long, laxly hairy in lower half; anthers about 3 mm long, 1.5 mm wide. Disk angled, pubescent on upper surface. Ovary oblong-cylindrical, 5-ribbed, with rounded top, about 7 mm long and 3 mm in diameter, densely silky-hairy and scaly, the hairs concealing the scales. Style short, stout, with abrupt transition to ovary, 5-6 mm long, 2.5 mm in diameter, lengthening to 15 mm in young fruit and falling before fruit matures, glabrous. Stigma 5-lobed, capitate. Young capsule 2.4 cm long, densely hairy and scaly.

Distribution. *R. impositum* has been recorded only from the Latimodjong mountain range in south-west central Sulawesi (Celebes).

Habitat. Open situations on ridges, 2400-3000 m.

Rhododendron impositum
CHRISTABEL KING

RHODODENDRON BROOKEANUM

Subgenus *Rhododendron*
Section *Vireya* subsection *Vireya* series *Javanica*

Rhododendron brookeanum is one of the larger-flowered species in the series *Javanica*. It was named after Sir James Brooke, who became Rajah of Sarawak in 1841. The species occurs, growing both epiphytically and terrestrially, in Sabah (300-1800 m) and in Sarawak. Argent notes that in Sarawak the species is found growing near sea-level on large trees and cliffs overhanging rivers, as well as (occasionally) in mangrove vegetation.

Rhododendron brookeanum was discovered and described by Hugh (later Sir Hugh) Low, who was so impressed with its beauty that he enthused:- 'I shall never forget the first discovery of this gorgeous plant; it was epiphytal upon a tree which was growing in the waters of a creek. The head of flowers was very large, arranged loosely, of the richest golden yellow, resplendent when in the sun; the habit was graceful, the leaves large.'

Thomas Lobb introduced the species to the Veitch Nurseries from whence material was sent to Kew in 1856. The Kew plant was figured for *Curtis's Botanical Magazine* by Fitch (tab. 4935) and published in September of that year. Mr and Mrs T.F. Allen collected several clones and introduced them into cultivation. Two of these, 'Mandarin', with large red flowers, and 'Raja' with scented yellow blooms, were awarded RHS First Class Certificates in 1970 and 1972 respectively.

The 'Raja' clone, later identified by Dr Sleumer as *R. brookeanum* var. *gracile*, grew at about 2400 m. In cultivation this clone flowers from March to May. The original plant from which another clone, 'Mesilau' was taken, grew in the lower montane zone, alongside 'Mandarin', on a granite boulder in the bed of Sungei Mesilau Kechil, a tributary of the Mesilau East River. This site is at an elevation of about 1450 m. The plant was, rather surprisingly, terrestrial (*R. brookeanum* is usually epiphytic at such altitudes). In cultivation 'Mesilau' flowers from October to January. Several other species from the series *Javanica* occur in Borneo, but Sleumer considers that *R. triumphans* from south Annam (Vietnam) is probably the closest relative of *R. brookeanum*. The latter is distinguished by its glabrous corolla and style, densely velutinous-hairy and scaly ovary, and medium-sized, oblong or narrowly elliptic leaves. It should be noted that Argent (1988) did not consider *R. brookeanum* to be distinct and reduced it to subspecific rank under *R. javanicum*.

It was not long after Thomas Lobb had introduced the species that hybridists found *R. brookeanum* to be an exciting and productive parent for their purposes. Now the species itself is once again enjoying the popularity it acquired in the late nineteenth century. Mr Allen has given details of cultivation and propagation methods for Sabah rhododendrons, in his paper in the *RHS Rhododendron and Camellia Yearbook* for 1971. In the case of *R. brookeanum*, cuttings can be rooted with the aid of hormones in a 50 per cent sand, 50 per cent peat mix with bottom heat, (late autumn seems to be the best time for rooting cuttings). Seeds germinate readily but are not so easily reared.

Rhododendron brookeanum Low ex Lindley in J. Hort. Soc. London 3: 82, *cum ic.* (1848).
Azalea brookeana (Low ex Lindley) O. Kuntze, Revis. Gen. Pl. 2: 387 (1891).

Description. An epiphytic or terrestrial shrub up to 2(-4)m in height, with fleshy roots and more or less sturdy, somewhat angular, scaly branchlets; internodes 2-15 cm long. Leaves persistent in indistinct pseudowhorls of 3 or 4, occasionally 5, at upper nodes, or spaced along the branches, shortly petiolate, thinly coriaceous, glossy, oblong-lanceolate or elliptic, 8-15(-28) cm long, 2.5-4(-9) cm wide, base cuneate or attenuate, sometimes rounded, apex acute to acuminate, usually cuneate to attenuate, rarely subtruncate or rounded, margin sometimes edged with purple, upper surface with caducous scales which leave dark pits, lower surface with more or less persistent scales, midrib pale and prominent, main nerves 12-14(-24) paired, other venation inconspicuous; scales minute, stellate. Inflorescence umbellate, with (5-)14(-18) lemon-scented flowers. Floral bud-scales caducous. Bracteoles 4-4.5 cm long, 1.5-2 cm wide, filiform to spathulate. Pedicels 3-4 cm long, 3 mm in diameter, sparsely scaly. Calyx an undulate, glabrous rim, 4-5 mm in diameter. Corolla with a golden yellow tube and bright orange-red lobes (other forms of the species can be orange to pink with white or cream centre, or pale yellow or entirely golden yellow), funnel-shaped, (5-)6-7(-8) cm long, lemon-scented; tube 2.5-3(-3.5) cm long, with 5 basal pouches, (6-)8-10 mm in diameter at base, widening upwards to 2-2.5 cm at throat, externally glabrous, minutely pubescent at base internally; lobes 5, oblong-obovate or suborbicular, about 4 cm long, 2.5-3(-3.5) cm wide. Stamens 10, subequal, not exserted; filaments finely pubescent in lowest third; anthers crimson to blackish, oblong, curved (4-)5-6 mm long. Disk lobed, pubescent on upper surface. Ovary more or less cylindrical, tapering into style or not, 8-10 mm long, (2-)3-4 mm in diameter, densely greyish silky-hairy and scaly, the scales often concealed by the hairs. Style pale greenish yellow, purple near apex, as long as stamens, 3-3.5 mm long, pubescent at base, glabrous above. Stigma crimson, capitate, 5-lobulate, about 5 mm in diameter, lobules emarginate. Capsule (described from collections in Kew herbarium) cylindrical to fusiform, somewhat angular, especially at base, splitting into recurving valves, (3.5-)5.7-7.2 cm long, 0.7-1.3 cm in diameter. Seeds 7-9 mm long including tails.

Distribution. Vars *brookeanum*, *cladotrichum* and *gracile* are confined to Borneo, var. *extraneum* comes from Sumatra.

Habitat. On rocks and mangroves, (near sea-level-)300-1525 (-1800) m.

Rhododendron brookeanum
WALTER HOOD FITCH

RHODODENDRON EDGEWORTHII

Subgenus *Rhododendron*
Section *Rhododendron* subsection *Edgeworthia*

Rhododendron edgeworthii was introduced by J.D. Hooker who described it in 1851 from a plant he had discovered in Sikkim in 1849. Hooker named it after his friend 'P. Edgeworth of the Bengal Civil Service', who at that time was Commissioner of Mooltan and an authority on the botany of north-west India. The plant described as *R. bullatum* from Delavay's 1886 collection in north-west Yunnan is now considered to be conspecific with *R. edgeworthii*, as is *R. sciaphilum*, which was described from a plant collected in east Burma some years later.

There have been many re-introductions of the variable *R. edgeworthii* by George Forrest (who first sent seeds in 1910), Farrer, Kingdon Ward, Rock, Ludlow & Sherriff, McLaren, Yü, and Cox & Hutchinson, which is not surprising, as this is one of the most beautiful members of subgenus *Rhododendron.* The large flowers, white or white flushed with rose-pink, borne on a relatively small bush, are reminiscent of subsection *Maddenia* in their lily-like corollas and are as deliciously fragrant. The leaves too are attractive, deep green and puckered above, tawny- to brown-woolly beneath. *Rhododendron edgeworthii* is not an easy species to grow successfully, being variably tender and very fussy as to drainage (Hooker found it growing happily on landslides — hardly an easy habitat to reproduce!). In glasshouse-grown plants, the delicate pink coloration tends to disappear and if grown in shade outside, the plant quickly becomes leggy. Nevertheless, it will flourish in a container, provided it is given perfect drainage so is well worth trying in the conservatory. And it can be grown successfully outside in the milder parts of the country; hardier forms such as Rock 59202, introduced in 1923, will sometimes survive in colder areas given adequate protection. A challenging species, but one which often rewards perseverance, and the beauty and fragrance of the species in full bloom in April, makes the extra time and trouble taken with it worthwhile.

Rhododendron edgeworthii is the type species of subsection *Edgeworthia* in Cullen's revision, which contains two additional species, *R. pendula* and *R. seinghkuense.* All three species possess the same dense indumentum of curled brownish or whitish hairs which characterizes the subsection and serves to distinguish it from all other groups. It is related to subsections *Maddenia, Moupinensia, Boothia* and *Camelliiflora.*

Confirming its merit (and popularity) are the four awards received by the species. Two Awards of Merit, one in 1923 for a plant shown as *R. bullatum* by T.H. Lowinsky, Sunninghill; the second in 1946, also shown as *R. bullatum*, this time by Lord Aberconway, Bodnant. The other two awards were First Class Certificates, the first in 1933 for a plant exhibited by Lt. Col. L.C.R. Messel, Nymans; the second in 1937, again for a plant shown as *R. bullatum*, by Lionel de Rothschild, Exbury. The accompanying illustration was painted by Walter Fitch from a plant growing at Kew, which flowered in May 1856.

Rhododendron edgeworthii Hook. fil., Rhodo. Sikkim Himal. t. 21 (1849).
R. bullatum Franchet, Bull. Soc. Bot. France 34: 281 (1887).
R. sciaphilum Balf. fil. & K. Ward, Notes Roy. Bot. Gard. Edinburgh 10: 146 (1917).

Description. An evergreen, epiphytic or terrestrial, scrambling shrub up to 2.5 m (occasionally to 3 m) in height; branches with a dense pale brown-orange indumentum of curled hairs. Leaves coriaceous. oblong-ovate, oblong or ovate-lanceolate or occasionally elliptic, (2.5-)6-15 cm long, 2-5(-6) cm wide with rounded or shortly tapered base and usually acuminate apex; upper surface very dark green, bullate, glabrous, lower surface densely tawny-woolly and sparsely scaly, the scales small, golden and concealed by the hairs. Petiole 6-20 mm long. Inflorescence terminal, with 2 or 3 flowers. Pedicels up to 2 cm long, densely shaggy. Flowers usually fragrant. Calyx large, leafy, 5-lobed, lobes unequal, oblong-orbicular, 1.1-1.4 cm long, externally densely shaggy and scaly and shaggy on margins, finely pubescent within. Corolla fleshy, funnel-campanulate, (3.5-)4.5-6(-7.5) cm long, white, sometimes flushed with pink, with or without a yellow blotch in the base of the tube; tube (1.9-)2.5-3(-4) cm long, externally scaly, glabrous within. Stamens 10, unequal, 1.5-5 cm long, declinate; filaments densely hairy towards base. Ovary densely woolly with lobed basal disk. Style declinate, more or less equalling corolla in length but exceeding stamens, hairy and/or scaly towards base. Stigma capitate, lobulate. Capsule densely woolly, oblong-globose, about 1.8 cm long in persistent calyx.

Distribution. This species has a very wide distribution over India (Arunachal Pradesh, west Bengal, Sikkim) Bhutan, east Burma and China (Yunnan, south Xizang).

Habitat. As an epiphyte in dense forest and as a terrestrial in scrub or on steep hillsides, 2100-3300(-3965) m.

Rhododendron edgeworthii
WALTER HOOD FITCH

RHODODENDRON DALHOUSIAE var. DALHOUSIAE

Subgenus *Rhododendron*
Section *Rhododendron* subsection *Maddenia*

Our frontispiece, reproduced from that published in Hooker's *Rhododendrons of the Sikkim Himalaya* illustrates *R. dalhousiae* growing epiphytically in its native country. It is a beautiful plant, producing large, strongly fragrant, lily-like flowers from April to July, and is hardy enough to be grown outside in the mildest parts of Britain, providing it is given adequate protection. For the rest of the country it is a species particularly suited to container-growing in the shelter of a glasshouse.

Rhododendron dalhousiae was discovered by J.D. Hooker in 1848 in Sikkim, growing at an altitude of between 1830 and 2898 m. It was described by him the following year when he named the species after Christina, Countess of Dalhousie (1786-1839) the wife of the Govenor-General of India. In 1850, Hooker introduced his new species into cultivation, sending seed to both Sunningdale and the Royal Botanic Garden in Edinburgh. Plants raised from this seed were offered for sale, unnamed, in 1851. The species flowered for the first time in cultivation in 1853 at Dysart House, Kirkaldy, Fife, having taken three years to reach that stage, after being grafted onto *R. ponticum* by the method known as inarching (see glossary). It was figured for *Curtis's Botanical Magazine* (tab. 4718) that same year by W.H. Fitch.

Rhododendron dalhousiae has won several awards from the RHS since its introduction (Award of Merit in 1930 and 1974, First Class Certificate in 1974) and has been successfully used as a parent of some notable hybrids. Its main claim to fame however, is as the pollen parent of the well-known lepidote-elepidote hybrid derived from a cross with *R. griersonianum* and named rather predictably 'Grierdal'. The species most closely related to *R. dalhousiae* is *R. lindleyi*, which is distinguished from the former by its altogether whiter flower and smaller, narrower, eciliate leaves. Cullen (1980) assigned both *R. lindleyi* and *R. dalhousiae* to subsection *Maddenia*, which itself is related to subsections *Edgeworthia* and *Triflora*. He recognised two varieties of *R. dalhousiae*, the type variety, described here, and var. *rhabdotum* (= *Rhododendron rhabdotum*) which differs from *R. dalhousiae* merely in flower colour (see plate 23).

Rhododendron dalhousiae Hook. fil., Rhodo. Sikkim Himal. t. 1 *non* t. 2 (1849), var. **dalhousiae**.

Description. Usually epiphytic, sometimes terrestrial, leggy, erect or sprawling shrub to 3 m in height with red-brown peeling bark; young growth loriform-setose. Leaves narrowly elliptic or obovate, (7.5-)10-17(-29) cm long, 3.5-7.5 cm wide, with tapered base and more or less rounded apex. Petioles 1-1.5 cm long, variably loriform-ciliate. Inflorescence of 2-3(-6), often fragrant flowers. Pedicels 1.5-2 cm long, scaly and pubescent, accrescent in fruit. Calyx leafy, conspicuous, with 5 oblong, rounded lobes 10-15 mm long, externally scaly at base, centre of lobes hairy externally. Corolla greenish in bud, opening white, cream or strongly tinged with green, often yellowish within, narrowly or more widely funnel-campanulate, 8.5-10.5 cm long, 5-6 cm across, sparsely scaly externally. Stamens 10, about 7 cm long; filaments pubescent in lower part. Ovary tapered into style, scaly. Style scaly towards base. Stigma capitate, lobulate. Capsule cylindrical to fusiform, 5-ribbed with persistent calyx and style base, 4-5.5 cm long, scaly. Seeds more or less fusiform, tailed at each end.

Distribution (of var. *dalhousiae*). Nepal, India (Sikkim, west Bengal), Bhutan and China (south Xizang).

Habitat. Growing epiphytically usually in forest and scrub, 1800-2450 m.

Rhododendron dalhousiae var. *dalhousiae*
WALTER HOOD FITCH

RHODODENDRON DALHOUSIAE var. RHABDOTUM

Subgenus *Rhododendron*

Section *Rhododendron* subsection *Maddenia*

As has been stated under *R. dalhousiae* var. *dalhousiae* (plate 22), Cullen has split this species into two varieties on the basis of corolla colour. Both are included here, not only for the beauty of the plants, but also for the contrasting styles of the two artists, both of whom have painted fine plates of this splendid species.

Originally ranked as a species itself, var. *rhabdotum* differs from var. *dalhousiae* in flower colour alone, the latter variety lacking the red stripes of the former, for which reason the rank of *R. rhabdotum* has been reduced. This would seem to be the best arrangement, as the two plants in other respects are so similar, but it is a little suspicious that no intermediate forms with faint or ill-defined stripes or any other indications of kinship with var. *dalhousiae* have been recorded and that, as Cox (1979) reports, var. *rhabdotum* comes true from hand-pollinated seed. *Rhododendron dalhousiae* var. *rhabdotum* was discovered by R.E. Cooper near Punakka in Bhutan in May 1915, growing on dry rock faces at an altitude of 2440 m. The specimen he collected was described by Bayley Balfour as *R. rhabdotum* in 1917, and published jointly with Cooper. It was not introduced into cultivation until February 1925, when Kingdon Ward collected seed near the Bhutan-Assam border on his way back from the Tsangpo gorge. Subsequently Ludlow and Sherriff re-introduced the plant from the same area, where it is apparently not uncommon. Var. *rhabdotum* flowered for the first time in England in August 1931 and it received an Award of Merit from the RHS when exhibited by Lady Aberconway and the Hon. H.D. McLaren, Bodnant, whose plants had been raised from Kingdon Ward's seed. A second award, in the form of a First Class Certificate, was given to the variety when Lionel de Rothschild, Exbury, exhibited it at the RHS in 1934. Both varieties of *R. dalhousiae* are tender except in a few of the mildest gardens of the British Isles, and even there some protection is advisable. Fortunately the species is late-flowering, var. *dalhousiae* coming into bloom about two months earlier than var. *rhabdotum* and lasting from April to July. Var. *rhabdotum* does not start to flower until June.

Tab. 9447 was painted by Lilian Snelling, from a plant grown out-of-doors at Minehead in Somerset by Mrs G. Blathwayt, where it flowered for the first time in 1933.

Rhododendron dalhousiae Hook. var. **rhabdotum** (Balf. fil. & Cooper) Cullen in Notes Roy. Bot. Gard. Edinburgh 36: 107 (1978).
R. rhabdotum Balf. fil. & Cooper in Notes Roy. Bot. Gard. Edinburgh 10: 141 (1917).

Description. As has been stated, the description of var. *rhabdotum* differs from that of var. *dalhousiae* in flower colour only; the corollas of var. *dalhousiae* usually have a greenish cast and always lack the external longitudinal red stripes of var. *rhabdotum*.

Distribution. North India (Arunachal Pradesh), Bhutan, China (south Xizang). The distribution of var. *rhabdotum* appears to be somewhat more restricted than that of var. *dalhousiae*.

Habitat. Dry rock faces, open hillsides or in forests where it grows both as an epiphyte and a terrestrial, 1500-2700 m.

Rhododendron dalhousiae var. *rhabdotum*
LILIAN SNELLING

RHODODENDRON FLETCHERIANUM

Subgenus *Rhododendron*
Section *Rhododendron* subsection *Maddenia*

Allied to *R. valentinianum*, this species differs in geographical range and the following characters: distinctly crenate margin around the upper half of the leaf, midrib usually not setose above, ovary conspicuously setose towards apex, the setae persisting in fruit, and, the most obvious difference of all, the sparsely scaly, green lower leaf surface which in *R. valentinianum* is brown with dense, overlapping scales. *Rhododendron fletcherianum*, which is known from two collections only, occurs in the forests of south-east Xizang at altitudes between 4000 and 4300 m, while *R. valentinianum* is native to north-east Burma and south-west Yunnan, where it grows in scrub or on cliffs and stony slopes at the lower altitudes of between 2700 and 3600 m. Many experts consider that *R. fletcherianum* resembles *R. ciliatum* more closely than it resembles *R. valentinianum*, and certainly the affinity between these taxa is strongly apparent. Both *R. fletcherianum* and *R. valentinianum* are yellow-flowered members of subsection *Maddenia* which is a predominantly white-flowered alliance: the flowers of the former species are paler in colour than those of the latter.

Rhododendron fletcherianum was discovered and introduced by Joseph Rock in 1932 (he collected seed late in that year, Rock 22302) and was known in cultivation as Rock's form of *R. valentinianum* until Davidian described it as a new species in 1961. It was named after H.R. Fletcher (1907-78), the Regius Keeper of the Royal Botanic Garden, Edinburgh from 1956 to 1970. Rock's plant was growing in forests of the alpine region of what was then Tibet, but is now called Xizang, on the Sola-La, Tsarong Province in the south-east of that country. It is one of the finest, small, yellow-flowered rhododendrons, a pretty and free-flowering species which is hardier than *R. valentinianum* and has been used successfully as a parent to produce dwarf yellow-flowered hybrids. It is easy to propagate from cuttings and seed, and flowers at an early age. The flowering period is from March to May and it will bloom again in autumn if summer rainfall is heavy. The early flowering period exposes the new growth to frost damage, so the species should be planted in a sheltered position, but one in which it receives plenty of light.

The accompanying plate (tab. 508 new series) was painted by Lilian Snelling in May 1951 from a piece of a plant growing in the Royal Botanic Garden in Edinburgh, which was sent to Kew by Sir William Wright Smith for that purpose.

A clone 'Yellow Bunting' with primrose-yellow flowers, exhibited at an RHS show in April 1964 by E.H.M. and P.A. Cox, Glendoick Gardens, Perth, received an Award of Merit.

Rhododendron fletcherianum Davidian in Rhododendron Camellia Year Book 16: 103 (1961).

Description. A broad, erect, evergreen shrub up to 1(-1.3) m in height, at first compact, later somewhat spreading, with reddish brown, peeling bark and scaly and densely loriform-setose young growth. Leaves oblong-elliptic or oblong-lanceolate, 2.3-5.6 cm long, 1.1-2.8(-3.1) cm wide, with cuneate, decurrent base, obtuse or acute apex and crenate, loriform-ciliate margins; upper surface dark green, midrib glabrous, lower surface green, sparsely scaly. Petiole 4-10 mm long, winged, densely loriform-setose. Inflorescence a terminal, compact truss of 2-5(-6) flowers. Pedicels sparsely scaly, densely loriform-setose, somewhat accrescent. Calyx leafy, with 5 oblong lobes, 8-10 mm long, scaly and loriform-ciliate. Corolla pale yellow, widely funnel-shaped, 3.6-4.2 cm long. Stamens 10, unequal, shorter than corolla; filaments hairy in lower half. Ovary densely scaly and densely setose towards apex, the setae persisting in fruit. Style impressed, variably setose in lower part. Capsule ovoid, 6-9 mm long, scaly and setose, with persistent calyx.

Distribution. China (Tsarong (Tsarung) province of south-east Xizang) *R. fletcherianum* vicariates with *R. valentinianum*. It is known only from Rock's two collections and is apparently confined to the forests of the province.

Habitat. Forests, 4000-4300 m.

Rhododendron fletcherianum
LILIAN SNELLING

RHODODENDRON FORMOSUM var. FORMOSUM

Subgenus *Rhododendron*
Section *Rhododendron* subsection *Maddenia*

Discovered by a 'Mr Smith' in 1815 on the Khasia hills ('bordering on Silhet') in Assam and described by Wallich in 1832, *R. formosum* is one of the hardiest members of subsection *Maddenia*; even so it is barely hardy enough for the very mildest and most sheltered of British gardens. Nevertheless it is a lovely species (the specific epithet means 'beautiful') and worth trying against a protective wall or in similarly sheltered situations, but it thrives best in a glasshouse where, planted in a container, its fragrance can be fully appreciated. The very dark green, neat fringed leaves make a perfect foil for the red-tipped buds and relatively large, lily-like, pink and white, often yellow-blotched flowers which are produced from April to June. The flowers vary somewhat in colour, but the most beautiful form is that shown in tab. 4457, painted by Fitch, material of which came from Syon garden.

In 1837, the plant was discovered on the Khasia Hills by J. Gibson, the then Duke of Devonshire's collector, and in 1841 Paxton described and figured it as *R. gibsoni*, but this name was reduced to synonymy by Hutchinson in 1919. Several years later in 1845, it was Gibson who introduced '*R. gibsoni*' into cultivation and plants flowered for the first time in the Duchess of Northumberland's glasshouse at Syon and at Kew in 1849, the latter plant having been presented to the Royal Gardens by Mr Low of Clapton. The species was collected and re-introduced by various other collectors over the years, and in 1960, a form bearing white, pink-tinged flowers with an orange throat, grown at the Royal Botanic Garden, Edinburgh, received an Award of Merit when exhibited at the RHS.

The three subseries of series *Maddenii* as classified by Hutchinson in *The Species of Rhododendron* (1930) have been re-classified by Cullen into four informal units comprising a total of 36 species in all. *Rhododendron formosum* is split into two varieties distinguished by leaf width and geographical range; var. *formosum*, with leaves 10-16 mm wide is confined to Meghalaya (India) and var. *inaequale* with leaves 15-21 mm wide occurs in Meghalaya, Manipur and Arunachal Pradesh.

Rhododendron formosum Wall., Plant. Asiat. Rar. 3(3): t. 207 (1832) var. **formosum**.
R. gibsoni Paxton's Mag. Bot. 8: t. 217 (1841).
R. formosum var. *salicifolium* C.B. Cl., Fl. Brit. India 3: 473 (1882).
R. iteaphyllum Hutch. in Notes Roy. Bot. Gard. Edinburgh 12: 83 (1919).

Description. An erect, evergreen, compact to lax shrub, up to 2(-3) m in height; young growth scaly and loriform-setose. Leaves coriaceous, narrowly elliptic to linear-elliptic or linear-obovate, (2.6-)4.2-7.6 cm long, 1-2.1(-2.4) cm wide, base tapered, decurrent, apex acute to acuminate; upper surface very dark green, not scaly at maturity, lower surface glaucous (whitish papillose) scaly, scales unequal, about their own diameter apart, leaf margins loriform-ciliate. Petiole 3-8 mm long, loriform-setose. Inflorescence a terminal cluster of 2 or 3 fragrant flowers; floral bud-scales more or less persistent during flowering. Pedicels 6-13 mm long, stout, scaly. Calyx an oblique, scaly, weakly loriform-ciliate disk. Corolla white or white flushed with pink, with or without a yellow basal blotch and/or narrow pink to crimson stripes running externally from base of tube to centre point of lobe-apex; funnel-shaped or widely so, 4-5(-6.3) cm long, externally scaly and pubescent at base. Stamens 10, unequal, 2.5-5.8 cm long; filaments pubescent towards base. Ovary scaly. Style impressed, slender, scaly for half or more of its length. Stigma capitate. Capsule about 1.6(-2) cm long, scaly.

Distribution. Var. *formosum* occurs only in India (Meghalaya).

Habitat. Hillsides among other ericaceous shrubs, (600-)1450-2300 m.

Rhododendron formosum var. *formosum*
WALTER HOOD FITCH

RHODODENDRON VEITCHIANUM

Subgenus *Rhododendron*
Section *Rhododendron* subsection *Maddenia*

W.H. Fitch excelled himself when he produced the beautiful accompanying plate of this splendid plant, and so successfully condensed a large-flowered species into a small space, that none of its impact was lost.

Rhododendron veitchianum was introduced into cultivation by Veitch's collector, Thomas Lobb, in about 1850. In recognition of the introduction, W.J. Hooker named the plant after the elder James Veitch, but also included James Veitch Jnr in the dedication. The species was figured for *Curtis's Botanical Magazine* in 1857 (tab. 4992), a plant having been exhibited by Veitch at an RHS show in May of that year. There, it attracted the attention of a writer for the *Gardeners' Chronicle* and gained mentions in the next two issues of that widely read journal.

Rhododendron veitchianum is a native of the mountains of central and southern Burma, Thailand and Laos and was first discovered by Thomas Lobb in the mountains east of Moulmain in Tenasserim on the Burmese coast, and it was from there that he sent the original material to Veitch in England. It is a plant of evergreen forests and open scrub, most often growing as an epiphyte, but sometimes terrestrial, or on moss-covered rocks, usually at elevations between 900 and 2400 m. Being tender, it makes a fine subject for the cool glasshouse, rarely exceeding 2 m or so in height in cultivation, although in the wild it can develop into a tree of double that stature. Huge white or pink-tinged fragrant blooms, sometimes with a yellow blotch in the throat, and in some forms with crisply frilled corolla-lobes, are freely produced and contrast with the comparatively small leaves, which are scaly and extremely glaucous beneath. Despite the fact that several distinct forms of varying appeal are in cultivation, this spectacular species received no awards until 1978 when the clone 'Margaret Mead' was exhibited at an RHS show by Geoffrey Gorer of Sunte House, Haywards Heath, Sussex and received an Award of Merit. That this is its only award seems particularly strange, considering the popularity of the species as a glasshouse plant in the nineteenth century, and the many re-introductions, the most recent of these being by P.G. Valder and Sir Peter Smithers.

One of the most closely related species to *R. veitchianum* is the familiar *R. formosum* (see Plate 25) which is, however, an altogether smaller plant.

Rhododendron veitchianum Hook. fil. in Curtis's Bot. Mag. 83: t. 4992 (1857).
R. formosum Wall. var. *veitchianum* (Hook. fil.) Kurz in J. Asiat. Soc. Bengal 46: 276 (1887).
R. cubittii Hutch. in Notes Roy. Bot. Gard. Edinburgh 12: 78 (1919).
R. smilesii Hutch., *op. cit.* 71.

Description. A terrestrial or epiphytic shrub up to 2(-3.7) m in height with reddish brown bark; young growth scaly and with some fugitive loriform hairs. Leaves coriaceous, obovate or elliptic, (5-)6.5-10 cm long, (0.2-)2.8-4.6 cm wide, tapering to the base and with shortly acuminate, mucronate apex, dark green and glabrous above, glaucous and sparsely scaly beneath; scales unequal, golden reddish brown. Petioles about 1 cm long. Inflorescence terminal, (1-)2-5-flowered; flowers fragrant. Pedicels stout, 3-8 mm long, scaly. Calyx oblique, almost obsolete, shallowly 5-lobed, scaly and sparsely ciliate. Corolla white, with or without a yellow blotch in the throat, sometimes tinged with pink or green, funnel-campanulate, 5-6(-7) cm long overall; tube 2.6-3.4(-4) cm long, sparsely pubescent at the base externally, somewhat scaly; lobes spreading, with or without exaggeratedly frilled or undulate margins. Stamens 10, unequal; filaments 3.5-5.8 cm long, pubescent in lower part. Ovary conical, tapering into the style, scaly, with lobed, glabrous basal disk. Style 5.5-6.8 cm long, the basal third to half scaly. Stigma capitate, 5-lobed. Capsule up to 3 cm long, ribbed, scaly, in persistent calyx.

Distribution. Burma, Laos and Thailand.

Habitat. Evergreen forests, open scrub, rocks and cliffs, 900-2400 m.

Rhododendron veitchianum
WALTER HOOD FITCH

RHODODENDRON MOUPINENSE

Subgenus *Rhododendron*
Section *Rhododendron* subsection *Moupinensia*

The charming rhododendron depicted in the accompanying plate is perhaps not as well known as its more famous hybrid, × 'Cilpinense' (*R. moupinense* × *R. ciliatum*) which, eclipsing its parent's other progeny, seems to have found its way into so many gardens. *Rhododendron moupinense* deserves more popularity and, indeed, produces much more colourful flowers than its pink-flowered offspring. It is a low and slow-growing plant with a dense, sturdy habit, small, neat glossy leaves and profuse, disproportionately large pink and/or white flowers. In the best forms the buds are deep pink, as are the top two or three corolla-lobes, the latter also spotted with crimson, and the bud-scales are tipped with scarlet. These forms should be diligently sought, being well worth the extra trouble. If early-flowering can be called a fault, then *R. moupinense* is guilty of this failing, blooming in February and March, when the flowers are frequently frosted. This should not deter the grower, for despite such setbacks, the plant can be depended upon to put on a good show each following year and will not always suffer damage if the frost is only a slight one. The species has compensating virtues — it is hardy, somewhat resistant to drought (it will grow happily in rock-gardens or old tree stumps) it flowers at the early age of 3-4 years, when only a few centimetres high, and is readily propagated from cuttings. All in all *R. moupinense* is a most desirable plant and should be represented in every collection.

Abbé David (Père David) discovered this plant in 1870, growing in the Moupine region of Sichuan in western China, and Franchet described the species some years later, in 1886, from David's specimens. It was found again in 1908, and this time introduced into cultivation by E.H. ('Chinese') Wilson. It was growing on Emei Shan (Mt Omei) and other nearby mountains. Wilson's seed was sent to the Arnold Arboretum in Massachusetts, where plants were raised from it and in spring 1909, one of these was sent to Kew where it was figured by Matilda Smith for *Curtis's Botanical Magazine* under tab. 8598. Wilson reported that the plant can attain a height of 1.3 m in the wild growing (usually) epiphytically on a selection of broad-leaved trees, but in cultivation and grown terrestrially it rarely exceeds 75 cm or so in height.

The species has received two Awards of Merit from the RHS since its introduction. The first in 1914, was for a white form exhibited by Ellen Willmott of Great Warley, Essex, and the other was awarded in 1937 to a plant with rose-pink, crimson-spotted flowers shown by Lord Aberconway of Bodnant.

The subsection *Moupinensia* comprises three species, all very similar, and is closely related to subsection *Maddenia*. *Rhododendron moupinense* is most closely allied to *R. dendrocharis* from which it differs in height, leaf size, calyx-lobe characters, corolla size and colour, and style length.

Rhododendron dendrocharis is a shrub growing up to 70 cm; its leaves do not usually exceed 17 × 10 mm, the glabrous calyx-lobes reach 3 mm in length, the corolla is smaller than that of *R. moupinense* (20-22 mm long) and the stamens exceed the style in length. The species is not in general cultivation.

Rhododendron moupinense Franchet in Bull. Soc. Bot. France 33: 233 (1886).

Description. Low, spreading, evergreen, frequently epiphytic shrub up to 1.5 m in height; young growth loriform-setose. Leaves coriaceous, glossy, convex, ovate-elliptic to elliptic, sometimes obovate, (25-)31-40(-46) mm long, (13-)16-25 mm wide, with a rounded or cordate base and obtuse apex, margins loriform-ciliate, upper surface pale green and glabrous with somewhat pubescent midrib; lower surface pale green or brownish, densely scaly. Petioles 3-9 mm long, scaly and loriform-setose. Inflorescence terminal, 1-2(-3)-flowered, the floral bud-scales scarlet at the apex and variably persistent during flowering. Pedicels 6-9 mm long, scaly and pubescent. Calyx 5-lobed, about 2 mm long, scaly and pubescent. Corolla white, variably flushed with pink and with crimson spots on upper lobes, openly funnel-campanulate, 30-38(-48) mm long, about 50 mm across, externally glabrous, densely pubescent within tube; tube 16-18 mm long. Stamens 10, unequal, 18-28 mm long. Ovary conoid, scaly. Style slender, declinate, exceeding stamens, glabrous or sparsely pubescent near base. Capsule oblong-cylindrical, 20-22 mm long, densely scaly.

Distribution. China, central Sichuan: (Cox (1985) records the species from Guizhou also).

Habitat. Rocks, cliffs and tree-stumps; most commonly growing epiphytically on broad-leaved trees, usually evergreen oak, 2000-4000 m.

Rhododendron moupinense
MATILDA SMITH

RHODODENDRON MONANTHUM

Subgenus *Rhododendron*
Section *Rhododendron* subsection *Monantha*

In the Bayley Balfour system of classification, *R. monanthum* was included in series *Boothii*, although acknowledged to be aberrant. Davidian placed it in his series *Uniflorum*. The best arrangement seems to be that put forward in Cullen's revision of the genus, where subsection *Monantha* (allied to subsections *Maddenia* and *Boothia*) has been created to include four closely related, but little-known taxa (*R. monanthum*, *R. flavantherum*, *R. kasoense* and *R. concinnoides*) which had formerly been placed in various other groups. These are linked by corolla shape, exserted stamens and style, and winged and finned seeds. Of the four, *R. flavantherum* and *R. kasoense*, previously members of series *Triflorum* are listed in the *RHS Rhododendron Handbook* (1980) as being in cultivation, while *R. concinnoides* was once in cultivation, but is no longer grown. In that same publication, *R. sulfureum* Franchet is mentioned, but this should not be confused with *R. sulfureum* Diels, which is synonymous with *R. monanthum*. It is *R. monanthum* which is depicted here, being the type species of the subsection, from which the other three species differ very little. Apparently *R. monanthum* has never been introduced into cultivation, which is a little surprising as it is widespread in the wild and although the flowers are small, they are colourful, being bright lemon yellow with red-brown conspicuous anthers. In its native habitat, the species is said to flower twice a year.

As *R. sulfureum* Diels, the species was first discovered and gathered by George Forrest in north-west Yunnan, in the Lupo Pass, but was later described as *R. monanthum* by Isaac Bayley Balfour and W.W. Smith, who recognised it as a new species. Kingdon Ward, Rock and Yü also found the plant in Yunnan and again in south-east Xizang, as did Forrest, while Farrer collected it in north-east Upper Burma. Although a terrestrial on open rocky slopes and in scrub, in preferred situations such as pine forest margins, this plant often grows epiphytically, not only on conifers, but on other trees too.

The accompanying line-drawing was prepared by the author from herbarium material of F 20879, which Forrest collected in October 1921 'on the Salwin-Kiu Chiang Divide' in Tsarong, south-east Xizang, at between 3600 and 4300 m. It was growing in scrub on rocky slopes and cliffs.

Rhododendron monanthum Balf. fil. & W.W. Sm. in Notes Roy. Bot. Gard. Edinburgh 9: 250 (1916).
R. sulfureum sensu Diels in Notes Roy. Bot. Gard. Edinburgh 7: 66 (1912), *non* Franchet (1887).

Description. A small, evergreen, often epiphytic, straggly, spreading shrub up to 1.2 m in height, with densely scaly young growth. Leaves coriaceous, ovate-elliptic or elliptic, (2-)3-4.5(-5) cm long, (1-)1.4-2.5 cm wide, with cuneate base and acute apex; upper surface dark green, scaly, scales usually persistent, lower surface glaucous, papillose, densely scaly, the scales unequal. Petioles 4-8 mm long, scaly. Inflorescence terminal, (1-)2-flowered. Pedicels 2-6 mm long, densely scaly. Calyx a minute, undulate, densely scaly rim. Corolla bright yellow, tubular-funnelform to tubular-campanulate, 1.4-2.3 cm long, tube 1-1.4 cm long, externally scaly; lobes scarcely spreading. Stamens 10, unequal, 1-2.2 cm long; filaments densely pubescent near base. Ovary scaly. Style exceeding the stamens, impressed, glabrous. Capsule oblong-cylindrical, 1.4-1.6 cm long, scaly.

Distribution. China (north-west Yunnan, southeast Xizang), north-east Upper Burma.

Habitat. Along forest margins where it grows epiphytically on conifers and other trees, or terrestrially at edges of thickets, amongst scrub or on open rocky slopes, 2450-3650 m.

Rhododendron monanthum
PAT HALLIDAY

RHODODENDRON YUNNANENSE

Subgenus *Rhododendron*
Section *Rhododendron* subsection *Triflora*

Rhododendron yunnanense is one of the most floriferous species in subsection *Triflora* and is ideal as a foil to the deeper tones of *R. trichanthum* and *R. augustinii*. It is completely hardy and the masses of white, pink or lavender flowers, speckled with yellow, green or red, resemble a swarm of butterflies. Blossom completely conceals the plant for a short period in May, after which the corollas carpet the ground in the manner of other members of the subsection *Triflora*. This latter characteristic is not always appreciated by gardeners, but the 'carpet' does not persist for very long, and indeed, the preceding show of flowers easily makes up for such a small inconvenience. The subsection is allied to subsections *Maddenia*, *Scabrifolia* and *Heliolepida*, perhaps to subsection *Lapponica* also, but more distantly so.

There are numerous unnamed forms of this variable shrub and these, planted in a group with other species of the subsection, produce a lovely display in pastel colours. Growing them in this manner helps to prevent the bark splitting, as it is prone to do after late frosts. Stock of *R. yunnanense* is easily increased by cuttings which root very readily and this species has the additional merit that when straggly, it can be pruned to advantage.

The species, named and described by Franchet in 1886, is another of Delavay's Chinese discoveries, and was introduced by him, from Houang-li-pin in western Yunnan, to Paris in 1889. The plants at Veitch's Coombe Wood nursery in 1894, are presumed to have been raised from Delavay's seed. It was one of these plants which Veitch sent to Kew in that year, and which flowered there in 1897. Material from it was used by Matilda Smith when she prepared tab. 7614 to accompany text supplied by J.D. Hooker, but for the present volume a new plate has been painted by Valerie Price, from a plant grown at Wakehurst Place, Sussex. Subsequent to Delavay's introduction, the familiar band of collectors active in China (Forrest, Kingdon Ward, Rock) gathered seed and sent it back to Britain, where the species soon became a favourite with gardeners.

A plant from Glasnevin Botanic Garden in Dublin, exhibited by Mr F.W. Moore, received an Award of Merit from the RHS in May 1904, and this was followed by a well-deserved Award of Garden Merit in 1934.

Rhododendron yunnanense Franchet in Bull. Soc. Bot. France 33: 232 (1886).
R. chartophyllum Franchet in Bull. Soc. Bot. France 33: 232 (1886).
R. chartophyllum forma *praecox* Diels in Notes Roy. Bot. Gard. Edinburgh 5: 217 (1912).
R. hormophorum Balf. fil. & Forr. in Notes Roy. Bot. Gard. Edinburgh 12: 117 (1920).
R. suberosum Balf. fil. & Forr., *op. cit.* 301.
R. aechmophyllum Balf. fil. & Forr., *loc. cit.* 13: 226 (1926).

Description. An erect, variably deciduous or evergreen, lax, often leggy, shrub up to 6 m in height; young growth scaly and sometimes loriform-setose, the setae soon caducous. Leaves chartaceous, elliptic or narrowly so, or oblanceolate, (2.5-)3.5-7(-10.4) cm long, (0.8-)1.2-2(-2.8) cm wide, with a cuneate base and acute, mucronate apex, usually glabrous above, with the exception of the midrib which is usually puberulent, laxly scaly and often somewhat glaucous beneath. Petioles 3-11 mm long, usually puberulent and more or less loriform-setose, at least when young. Inflorescence lax, terminal or terminal and axillary, each inflorescence-bud 3-5-flowered. Pedicels (5-)11-20 mm long, laxly scaly. Calyx an undulate, scaly and ciliate or occasionally glabrous rim, 0.5-1 mm long. Corolla white, pink or lavender, usually heavily spotted with yellow, green or red, open funnel-shaped, 6-lobed, somewhat zygomorphic, (1.8-)2.5-3.4 cm long, 2-4 cm across the lobes, more or less free of scales externally, pubescent internally. Stamens 10, unequal, long-exserted from throat, almost as long as corolla-lobes, 1.4-4 cm long; filaments pubescent below the middle. Ovary scaly with hairy basal disk. Style slender, exceeding stamens, glabrous. Stigma capitellate. Capsule 1.2-1.7(-2) cm long, scaly.

Distribution. China (Yunnan, Sichuan, Guizhou and possibly south-east Xizang), north-east Burma.

Habitat. Grows in a wide variety of habitats, from open rocky slopes to shady thickets and forest margins, and in dry or moist situations, 2100-4270 m.

Rhododendron yunnanense
VALERIE PRICE

RHODODENDRON TRICHANTHUM

Subgenus *Rhododendron*
Section *Rhododendron* subsection *Triflora*

This beautiful member of subsection *Triflora* is a very close relative of *R. concinnum* and is notable not only for the plethora of long bristling hairs which clothe almost the entire plant, but also for the deep red-purple flower colour of the best forms. When backlit by the sun, the plentiful inflorescences glow with colour and make such a spectacle that it is easy to forget the 10 to 12 years taken by the plant to reach the flowering stage. Care should be taken when selecting plants of this species for the garden, as some forms produce pale or dingy flowers. It is therefore best to buy plants in bloom, which means during late May and June, this being the latest flowering species in the subsection. To be enjoyed at their best, these slender trees or shrubs should be grown in a mixed group of species chosen from the *Triflora* subsection, which in flower colour vary from white through pink to deep blue- and deep red-purples. All are so floriferous that the resulting explosion of colour, albeit mainly pastel shades, completely conceals the relatively small leaves. A rich foil for the deep red-purple of *R. trichanthum* is the glorious 'blue' form of *R. augustinii*. 'Electra', another close relative, is a lighter blue-lavender cultivar of the latter species, with a large yellow-green flare in the throat. It, too, is a lovely plant worthy of inclusion in any collection.

Rhododendron trichanthum is a common species in the wild and forms dense thickets, particularly in the woodlands of west and south-west Sichuan, at altitudes of up to 3650 m. It was discovered on Wa Shan in that province by E.H. Wilson in 1903. He gathered seed the following year, when Messrs Veitch introduced the species into cultivation. Wilson subsequently collected seed for the Arnold Arboretum in 1908 and the species was re-introduced that year and again in 1910. Various other collectors have also found the species in Sichuan.

The original plate was painted in Edinburgh by Lilian Snelling for Sir Isaac Bayley Balfour sometime between 1916 and 1920. Prints were made and two were passed to a 'Mrs Ely' for 'trial colourings'. The illustration reproduced here is of one of them. *Rhododendron trichanthum* received an Award of Merit from the RHS when the clone 'Honey Wood' was shown in June 1971 by Major A.E. Hardy of Sandling Park, Kent.

Rhododendron trichanthum Rehder in J. Arnold Arb. 26: 480 (1945).
R. villosum Hemsley & Wilson in Kew Bull. 1910: 119 (1910), *non* Roth (1807).

Description. An evergreen shrub up to 3(-6) m in height; branches and young growth densely loriform-setose. Leaves ovate-elliptic or lanceolate, (4-)6-8(-11) cm long, (1.5-)2.8-3.7 cm wide, with cuneate to rounded base and acute to acuminate, mucronate apex, glabrous to loriform-setose and sparsely scaly above, pilose (at least on midrib) and laxly brown-scaly beneath. Petioles 4-10 mm long, scaly and densely pilose. Inflorescence terminal, 2-3(-5)-flowered. Pedicels (10-)12-17 mm long, scaly and densely loriform-setose, the setae twisted. Calyx undulate or 5-lobed, the lobes 1-2 mm long, externally scaly and loriform-setose. Corolla light mauve to dark red-purple, open funnel-shaped, somewhat zygomorphic, 2.8-3.8 cm long; tube short, 1.4-1.6 cm long, externally scaly and with flattened, loriform setae. Stamens 10, unequal, long-exserted from throat, but not exceeding the corolla-lobes, 2-3.3 cm long; filaments pubescent towards base. Ovary scaly and variably pilose. Style slender, exceeding stamens, usually glabrous. Capsule oblong-cylindrical, up to 1.9 cm long, scaly and variably bristly.

Distribution. China (west Sichuan)

Habitat. Thickets and woodland, 2300-3650 m.

Rhododendron trichanthum
LILIAN SNELLING

RHODODENDRON RACEMOSUM

Subgenus *Rhododendron*
Section *Rhododendron* subsection *Scabrifolia*

This is a familiar, very variable, easy to grow species and a good subject for a rhododendron-growing beginner. It can readily be raised from seed and on occasion has been used as a hedging plant, although its rather loose habit is not ideally suited for such purposes.

Originally classified as a member of series *Virgatum*, *R. racemosum* has been re-classified in the Edinburgh revision and is now placed in subsection *Scabrifolia*. The species is most closely related to *R. hemitrichotum* in the same subsection, which exhibits similar white-papillose lower leaf surfaces. There are several variants, none of which is worthy of formal recognition, however horticulturally distinct it is deemed to be.

Rhododendron racemosum was named and described by Franchet, having been discovered by the Abbé Delavay in April 1884 in the mountains of Yunnan, between 2400 and 3300 m where it is extremely plentiful. Delavay sent seed to the Jardin des Plantes in Paris and some of the resulting plants raised there in 1889 found their way to Veitch. In the same year the species reached Kew. Most stock now in cultivation, however, derives from more recent gatherings than those of Delavay.

Rhododendron racemosum is an attractive plant, very free-flowering, the top 12 cm of growth often smothered with blossom from March to May. The plants show to best advantage when grown in a group, comprising both dark and light forms, under taller rhododendrons. If space is limited, then there is one dwarf, compact form with deep cerise-pink flowers which is particularly desirable for the garden. It was collected by George Forrest in 1921 in the Sungkwei Pass, Yunnan (F. 19404) and is now commercially available. All forms will withstand fairly dry conditions and are not fussy as regards soil.

Some *R. racemosum* hybrids also are of interest, especially those raised from crosses with *Obtusum* azaleas which cast a questionable light on the relationships of the species.

Although the illustration reproduced here is a recent painting by Valerie Price, *R. racemosum* was figured for *Curtis's Botanical Magazine*, many years ago, by Matilda Smith (tab. 7301) when a plant flowered at Kew in March 1893. The specimen was sent to Kew from the Royal Botanic Garden, Edinburgh, by Bayley Balfour. On that occasion J.D. Hooker wrote the accompanying text and he comments in his account that plants growing at Kew and at Veitch's Coombe Wood Nurseries, flowered at about the same time. He also records that the flowers were very sweetly scented, but not all forms of the species are fragrant, any more than all the forms are completely hardy. The species has received several awards from the RHS, including a First Class Certificate in 1892 (Veitch & Sons, Chelsea), an Award of Merit in 1970 (Hydon Nurseries, Godalming) followed by another similar award in 1974 to a plant shown from Glendoick Gardens Ltd, Perth. It also received an Award of Garden Merit in 1930.

Rhododendron racemosum Franchet in Bull. Soc. Bot. France 33: 235 (1886).
R. motsuense Lév. in Feddes Repert. 13: 148 (1913).
R. iochanense Lév., *nom. nud.* (reference not traced).

Description. A variable evergreen shrub up to 3(-4.5) m in height; young growth often red or stained with red, scaly and sometimes puberulous. Leaves obovate or oblong-elliptic, (10-)15-54 mm long, (4-)7-30 mm wide, with obtuse base and obtuse or rounded, mucronate apex, glabrous above with sparsely pubescent midrib, densely scaly and white-glaucous, papillose beneath. Petioles 2-6 mm long, scaly, sometimes also puberulous. Inflorescences appearing terminal but axillary in uppermost leaf axils, often extending along 15 cm or more of the branch, each cluster 2-3(-6)-flowered. Bud-scales more or less persistent. Pedicels up to 17 mm long, scaly or minutely hairy. Calyx a densely scaly, undulate or shallowly lobed rim, 0.5 mm long. Corolla white or pale pink to deep pink, widely funnel-shaped, 5-lobed, 7-17(-23) mm long; tube 3.5-8.5 mm long, shorter than the lobes, somewhat puberulous within. Stamens 10, unequal, exserted, 8-23 mm long; filaments sparsely hairy near base. Ovary densely scaly. Style equalling or slightly exceeding the stamens, glabrous. Capsule (4-)7-10 mm long, scaly.

Distribution. China. (north, north-west, west and central Yunnan, south-west Sichuan).

Habitat. Many different habitats, usually in open situations and often covering large areas of steep hillsides with varied shades of pink blossom, (800-)2750-4300 m.

Rhododendron racemosum
VALERIE PRICE

RHODODENDRON SPINULIFERUM

Subgenus *Rhododendron*
Section *Rhododendron* subsection *Scabrifolia*

Rhododendron enthusiasts may look a little askance when they first see this oddly un-rhododendron-like plant. The curious tubular flowers with their tufts of exserted stamens, and the small wrinkled leaves are not what one expects of the genus. Equally unexpectedly, the species has close affinities with and fits fairly comfortably into subsection *Scabrifolia*, confirming its relationship by hybridizing in the wild (in the north of its range) with *R. scabrifolium*. The hybrid is known as *R. duclouxii* Lév. The corollas are filled with a watery nectar which raises the question of pollinators and fertilization; the latter is presumably a specialized affair, which does not seem to have excited any interest amongst workers on the genus.

Although not a particularly showy plant, *R. spinuliferum* adds interest to a collection. It is free-flowering and the scarlet corollas of the better forms, with their exserted black-and-white stamens contrast advantageously with the clear deep green of the foliage. Most forms have proved to be hardy despite early doubts, but the hardiness is admittedly variable, so some protection from frost is recommended. The species was first discovered by Abbé Delavay in 1891, above Tonghay in China, but it owes its introduction into European collections to Maurice de Vilmorin of Les Barres. Seeds of the species were sent by M. Ducloux to Vilmorin in 1907, from which plants were raised. One of these flowered for the first time at Les Barres in spring 1910 whence it was sent to Kew in the same year, to provide material for the preparation of tab. 8408. The resulting plate by Matilda Smith is not reproduced here. Instead, a new painting by Valerie Price is included. Rock and McLaren respectively re-introduced the species in 1932. In the wild *R. spinuliferum* flowers in April and May.

Rhododendron spinuliferum has received two awards from the RHS. An Award of Merit was given to a clone called 'Jack Hext', exhibited by Nigel T. Holman, Chyverton, Truro, Cornwall, in 1974. Another clone, 'Blackwater', from Brodick Castle gardens, Scotland, received a similar award when exhibited in 1977.

Rhododendron spinuliferum Franchet in J. de Bot. 9: 399 (1895).
R. fuchsiiflorum Lév. in Feddes Repert. 12: 284 (1913).

Description. An evergreen, often straggly shrub up to 3(-4.5) m in height with long, slender shoots; young growth grey-tomentose and setose, soon glabrous. Leaves coriaceous, glossy dark green, lanceolate-oblanceolate or obovate, convex (larger than those of *R. scabrifolium*), 2-4(-9.5) cm long, 1(-4.5) cm wide, with cuneate base and acuminate, mucronate apex, margins usually recurved, bullate and sparsely bristly above, finally glabrous except for persistent pubescence along the mibrib, strongly reticulate, lower surface sparsely scaly and hairy. Petioles 4-6(-10) mm long. Inflorescences terminal and axillary in uppermost leaf-axils, umbellate, (1-)2-3(-5)-flowered; floral bud-scales caducous or persistent. Pedicels 6-8(-12) mm long, tomentose. Calyx obsolete, rim-like, undulate or shallowly 5-lobed, densely tomentose and long-ciliate. Corolla erect, usually orange-red, sometimes pink, scarlet or crimson, tubular to narrowly funnel-shaped, tapering to both apex and base, (1.4-)1.7-2.5 cm long, filled with watery nectar, glabrous; lobes 5, erect or somewhat incurved, imbricate, about 8 mm long. Stamens l0, unequal, 1.5-2 cm long, bunched, exserted; filaments glabrous. Ovary conical, scaly and hairy, the scales concealed by the hairs, with a basal, lobed, glabrous disk. Style slender, straight, exserted, equalling or slightly exceeding the stamens, sparsely pubescent at base. Stigma capitate, lobulate, as wide as style. Capsule oblong or somewhat tapering upwards, 1.1-1.7 cm long, scaly and hairy.

Distribution. China (central and south Yunnan).

Habitat. Shady thickets and scrub, 800-2500 m.

Rhododendron spinuliferum
VALERIE PRICE

RHODODENDRON RUBIGINOSUM

Subgenus *Rhododendron*
Section *Rhododendron* subsection *Heliolepida*

Easily recognizable horticulturally, '*R. desquamatum*' is not so easily separated taxonomically from the admittedly variable *R. rubiginosum*. The species as a whole is very vigorous and hardy, and according to Cox is wind-hardy too, nor is it as fussy about soil, as are some species. If grown as a tree rather than a shrub, it makes a most attractive centre-piece in the garden. The form of *R rubiginosum* described here (previously known as *R. desquamatum*) is a particularly lovely rhododendron. The large, widely funnel-shaped, clear lavender, reddish mauve (or even purple) speckled flowers are produced in profusion in March and April; in fairness it must be said that the typical form of *R. rubiginosum* is equally floriferous, but the individual flowers are smaller and not so attractively coloured as those of the plant depicted in tab. 9497. *Rhododendron rubiginosum* was described in 1887 by Franchet from material collected the previous year in Yunnan by Père Delavay. The species was introduced to Paris by Delavay himself in 1889. Farrer, Rock and Yü also collected the species and Forrest sent seed back several times after his initial introduction, from both north-west Yunnan and north-east Upper Burma.

The plant illustrated by Lilian Snelling in tab. 9497 is of a different, more spectacular form which was probably first collected by Abbé R.P. Soulié at Tsekou in 1895. However, this showy form was not introduced until 1917, when George Forrest found it on the Shweli-Salween divide in west Yunnan. It was described as a new species, *Rhododendron desquamatum*, in 1920 by Bayley Balfour.

Rhododendron rubiginosum and *R. desquamatum* were maintained as separate species until Cullen reduced *R. desquamatum* to synonymy under *R. rubiginosum* in his 1980 revision where he noted that subsection *Heliolepida* is closely related to subsection *Triflora*. Not surprisingly, the species (as *R. desquamatum*) received an Award of Merit from the RHS when shown by Captain Talbot Fletcher, Margam Castle, Port Talbot, S. Wales, in 1938. He exhibited sprays from five different plants, raised from seed of Forrest's no. 24535 which admirably demonstrated the variability of the species, the flower colour ranging from clear pale mauve to reddish mauve. A clone of *R. rubiginosum* called 'Wakehurst' with large, tight trusses of 'Mallow Purple' (HCC 630/2) spotted flowers also received an Award of Merit from the RHS in 1960, when exhibited by Sir Henry Price of Wakehurst Place, Sussex.

Rhododendron rubiginosum Franchet in Bull. Soc. Bot. France 34: 282 (1887).
R. leclerei Lév. in Feddes Repert. 12: 284 (1913).
R. catapastum Balf. fil. & Forr. in Notes Roy. Bot. Gard. Edinburgh 13: 36 (1920).
R. desquamatum Balf. fil. & Forr., *op. cit.* 40.
R. stenoplastum Balf. fil. & Forr., *op. cit.* 60.
R. leprosum Balf. fil., *nom. nud.* (reference not traced).
R. squarrosum Balf. fil., *nom. nud.*, Notes Roy. Bot. Gard. Edinburgh 17: 266 (1930).

Description. An erect or spreading, evergreen shrub or small tree up to 10 m in height; young growth purplish, scaly. Leaves narrowly elliptic to elliptic or lanceolate, (4-)6-11.5(-12.5) cm long, (1.2-)2-4.5 cm wide, with cuneate base and acute to acuminate apex, upper surface glabrous, lower surface pale to dark brown, densely scaly, the scales contiguous or overlapping, flaky, unequal in size, often the larger scales darker in colour. Petioles 1-1.3 cm long, scaly. Inflorescence terminal, 4-8(-10)-flowered. Pedicels slender, 1.3-2.5(-3) cm long, scaly. Calyx an undulate rim, scaly. Corolla pink, pale to deep or reddish mauve or purple, occasionally white flushed with pink, speckled red, crimson, purple or brown, widely funnel-shaped, (1.5-)2-3(-3.8) cm long; tube (1.1-)1.5-2(-2.3) cm long, externally scaly, hairy within; lobes oblong, often with undulate margins. Stamens 10, unequal, 1.3-3.2 cm long; filaments pubescent below the middle. Ovary densely scaly with a lobed, glabrous basal disk. Style slender, declinate, exceeding longest stamens, densely scaly. Stigma capitate, lobulate. Capsule cylindrical, 1.1-1.6 cm long, scaly.

Distribution. China (north, north-west, west & central Yunnan, south-west Sichuan, SE Xizang), north-east Upper Burma.

Habitat. Forests, thickets, cane brakes and similar situations, 2500-3500 m.

Rhododendron rubiginosum
LILIAN SNELLING

RHODODENDRON MINUS var. MINUS

Subgenus *Rhododendron*
Section *Rhododendron* subsection *Caroliniana*

One of the few American rhododendrons and, with the exception of *R. lapponicum*, the only lepidote species to occur in the New World, this variable species is native and endemic to Tennessee, North and South Carolina, Georgia and Alabama, where it grows on the lower mountain slopes and plains. *Rhododendron minus* is not a spectacular shrub, partly because the young foliage develops at the same time as the flowers and overtops the inflorescences, tending to obscure them. It is, however, free-flowering from an early age and deserves a place in a woodland setting, where it is shown to best advantage, especially in the autumn when the foliage turns bright red before the leaves fall. It is this species which most often represents subsection *Caroliniana* in British gardens. A yellow-blotched, white-flowered variety with narrower leaves which was first introduced into cultivation in 1895 is available from nurseries and garden centres, as well as forms with pink, mauve and rosy purple flowers. Although extremely hardy in the south-eastern USA, in Britain it is wise to provide some protection from early spring frosts, as well as very sharp drainage, if any of the species in this subsection are to flourish. *Rhododendron minus* has now become a useful parent, for its hardiness and late-flowering period are both desirable characteristics which are freely passed on to its progeny, but plants are best raised from seed because cuttings do not always root readily. A certain amount of confusion surrounds the name of this plant.

Introduced originally as *R. punctatum* by John Fraser in 1786, Michaux first described the species as *R. minus* in 1792; but then the name *R. punctatum* was used again by Andrews in 1798 to describe a cultivated plant. In 1811 *R. minus* was introduced into England as *R carolinianum* and was re-introduced under that name from H.P. Kelsey's nursery in Carolina. Some years later, in 1902, Small published the name *R. cuthbertii* for a plant from Georgia. Then Rehder, in 1912, published a description of *R. carolinianum*, and this became the most familiar name for the species. It has since been reduced to synonymy under *R. minus* var. *minus*, which, according to the rules of the ICBN, is the correct and valid name for the taxon, since the traditional points of distinction between *R. carolinianum* and *R. minus* have proved variable and unreliable. In the Bayley Balfour system of classification, Rehder's *R. carolinianum* belongs to series *Carolinianum*, a group of three species. The series in its entirety has become subsection *Caroliniana* of section *Rhododendron* in the 1980 Edinburgh revision, a subsection which is closely allied to subsection *Heliolepida*. Following work by Duncan & Pullen (1962), the number of species has been reduced to one, *R. minus* comprising two varieties, (var. *minus* and var. *chapmanii*) which are distinguished by leaf apex and branch characters. The type variety, under discussion here, has acute-acuminate leaf apices, the branches are usually neither erect nor rigid and it is later-flowering than var. *chapmanii*. The subsection is itself closely related to subsection *Heliolepida*, but is considered distinct by reason of the differing geographical distributions, subsection *Heliolepida* being confined to the Old World (China and Burma) whereas subsection *Caroliniana* is confined to the New World (south-east USA).

As *R. punctatum*, tab. 2285 was painted for *Curtis's Botanical Magazine* in June 1821 by J. Curtis, from material supplied by John Walker.

The species received an Award of Merit from the RHS in 1968 when Colonel N.R. Colville of Launceston, Cornwall, exhibited a form with rosy purple flowers.

Rhododendron minus Michaux in J. Hist. Nat. 1: 412 (1792), var. **minus**.
R. punctatum Andr., Bot. Rep. 1: t. 36 (1798).
R. punctatum var. β Ker, Bot. Reg. 1: t. 37 (1815).
R. cuthbertii Small in Torreya 2: 9 (1902).
R. carolinianum Rehder in Rhodora 14: 99 (1912).

Description. An evergreen, compact or leggy, scaly shrub up to 3 m in height in cultivation, but occasionally up to 5 or 10 m in the wild; young shoots frequently stained purple. Leaves obovate, elliptic or broadly so, (1-)5.5-8(-11.4) cm long, (1.8-)2.5-3.5(-6.5) cm wide, base tapering, apex acute to acuminate; upper surface sparsely scaly, lower surface densely brownish-scaly, midrib pubescent on upper surface. Petiole 1-1.5(-2) cm long, scaly. Inflorescence a terminal, dense, (4-)5-8(-12)-flowered cluster, sometimes with several terminal flower-buds. Pedicels (8-)10-16 mm long. Calyx 5-lobed, lobes usually unequal, 0.5-2 mm long, scaly and sparsely loriform-ciliate. Corolla mauve, rosy purple or white, sometimes with faint spots, narrowly to widely funnel-shaped, (2.1-)2.5-3(-3.8) cm long; tube (9-)10-14(-22) mm long, sparsely pubescent within; lobes sparsely scaly externally, sometimes also sparsely pubescent. Stamens 10, slightly exceeding corolla-tube, 1.3-3 cm long; filaments densely pubescent towards base. Ovary densely scaly. Style slender, glabrous, occasionally with a few scales at base, becoming declinate. Capsule oblong-cylindric, 7-10(-13) mm long, densely scaly and with persistent calyx.

Distribution. South-eastern USA (Tennessee, North and South Carolina, Georgia and Alabama).

Habitat. Hillsides and open plains. Altitudinal range has apparently not been recorded.

Rhododendron minus var. *minus*
JOHN CURTIS

RHODODENDRON HIPPOPHAEOIDES

Subgenus *Rhododendron*
Section *Rhododendron* subsection *Lapponica*

In its native regions of Yunnan and Sichuan this attractive plant spreads over vast tracts of land forming an ankle-deep carpet, which when the plants bloom in May, resembles a pink, lavender, blue and purple sea and is reminiscent of the heather moors in Britain. In May the snows are melting and ample moisture is available, but later in the season, the land becomes very dry and the vegetation must contend with drought. Such conditions recommend the species as an ideal and hardy subject for British gardens, and indeed it is. *Rhododendron hippophaeoides* is one of the most popular and widely cultivated species of subsection *Lapponica*, easy to propagate from seed and cuttings. Although happiest when grown in sandy peat and disliking wet conditions in cultivation, it is known to occur in boggy ground in the wild and cultivated plants will usually survive in ground which remains wet over the winter. Flowers are very freely produced from March to May and are rarely damaged by frost. Forms vary, particularly in height, which can be as much as 1.5 m, so should be carefully selected for both height and flower-colour. All in all, *R. hippophaeoides* is a most desirable plant, being particularly attractive when planted in a group of mixed heights and flower colours. The species was discovered in flower by Frank Kingdon Ward in the spring of 1913 near the Yangtse bend in north-west Yunnan when characteristically, he enthused about the 'Lapponicum sea'. But the first introduction was by George Forrest who, returning in the autumn to the same general area where he had seen it in flower earlier in 1913, collected seed. It was re-introduced many times subsequently by various other collectors as well as by Forrest himself.

The species was given an Award of Garden Merit in 1925 and a plant with lavender-blue flowers exhibited by Lady Aberconway and the Hon. H.D. McLaren of Bodnant in 1927, received an Award of Merit from the RHS.

Subsection *Lapponica* is based on Balfour's series *Lapponicum* and is related to subsections *Triflora*, *Heliolepida* and *Rhododendron*. The species comprises two varieties distinguishable by style length and geographical range; var. *hippophaeoides* has a style-length of 4-10.5 mm; the styles of var. *occidentale* are longer being 13-16 mm in length. The latter variety does not occur in Sichuan. Both varieties hybridize in the wild.

Lilian Snelling painted tab. 9156, illustrating three colour forms, in March 1926.

Rhododendron hippophaeoides Balf. fil. & W.W. Sm. in Notes Roy. Bot. Gard. Edinburgh 9: 236 (1916).
R. fimbriatum Hutch. in Gard. Chron. 91: 438 (1932).

Description. A small, erect or sometimes sprawling, evergreen, scaly shrub up to 1.5 m in height, with slender, scaly branches. Leaves dark green, elliptic-oblong, 0.8-3.8 cm long, with cuneate base and rounded, obtuse apex, laxly scaly above, the lower surface yellowish buff with dense, overlapping scales. Petioles 2-5 mm long, scaly. Inflorescence a terminal, 4-7-flowered umbellate raceme. Pedicels 2.5-7 mm long, scaly. Calyx up to 1.8 mm long, the lobes often unequal, variably scaly and ciliate. Corolla bright rose or lavender to bluish purple, occasionally white, widely funnelform, 10.5-15 mm long, with very short, wide tube, pubescent within. Stamens 10, shorter than corolla; filaments pubescent towards base. Ovary scaly. Style usually red, shorter or longer than stamens, glabrous or sometimes pubescent at base. Capsule narrowly ovoid, 5-6 mm long, scaly.

Distribution. China (Yunnan, south-west Sichuan).

Habitat. Open slopes, often in marshy ground, 2400-4800 m.

Rhododendron hippophaeoides
LILIAN SNELLING

RHODODENDRON RUSSATUM

Subgenus *Rhododendron*
Section *Rhododendron* subsection *Lapponica*

The best forms of this beautiful hardy dwarf produce a profusion of intense blue-violet, open funnel-shaped flowers, in a genus where 'true blues' are non-existent. The flower colour is further enhanced by the contrasting tuft of snow-white hairs which fill, and are exserted from, the throat. Although the deeper colours are to be preferred, all forms of the species, even leggy plants, are worth growing, and are particularly effective planted in a group, when the range of flower colour from white and pale blue-lavender to indigo-blue or blue-purple will produce a glorious display.

George Forrest first collected and introduced *R. russatum* into cultivation in 1917 (as no. 13915). The plant was growing in open, moist, stony pastureland in the Kari pass in north-west Yunnan, at an altitude of 4000 m where it formed a shrub 1 m in height. It was flowering when gathered in June, but in cultivation the flowers usually appear a month or so earlier, from April to May, according to prevailing conditions. Forrest subsequently re-introduced this attractive species many times, as did Joseph Rock. *Rhododendron russatum* hybridizes naturally with the closely related *R. rupicola* and some of Forrest's (and Rock's) collections of *R. russatum* were later found to be crosses between these two species. Contrarily, a plant said to be raised from Forrest's number 16583 was initially named and described as *R. cantabile* by Bayley Balfour in 1922. As *R. cantabile*, the plant was figured in *Curtis's Botanical Magazine* under tab. 8963, from material supplied by Mr J.C. Williams of Caerhays Castle, Cornwall. At that point Mr Williams' specimen was recognized as being, without doubt, a form of *R. russatum*, and *R. cantabile* was promptly reduced to synonymy under *R. russatum*. Tab. 161, reproduced here, was prepared for the *Kew Magazine* by Valerie Price in spring 1989, from a plant growing at Wakehurst Place, Ardingly, Sussex. A form of this species with flowers of deep violet-blue exhibited at an RHS show in 1927 by A.M. Williams of Werrington Park, Cornwall, was given an Award of Merit, and in 1933, Lionel de Rothschild's plant from Exbury, with intense purple flowers, was awarded a First Class Certificate, the same plant when shown five years later in 1938, being given an Award of Merit.

Rhododendron russatum Balf. fil. & Forr. in Notes Roy. Bot. Gard. Edinburgh 11: 126 (1919).
R. cantabile Balf. fil. ex Hutch. in Curtis's Bot. Mag. 148: t. 8963 (1922).
R. osmerum Balf. fil. & Forr. in Stevenson, J.B., (ed.), Sp. Rhodod., 425 (1930), in synon., *nom. nud.*

Description. A sprawling, cushion-forming or erect evergreen shrub up to 1.8 m in height. Leaves narrowly to broadly elliptic or oblong, (6-)16-40(-65) mm long, 3-12(-23) mm wide, base cuneate, apex obtuse or rounded, mucronate, dark, matt green above, yellow or pale brown to rusty or red-brown beneath and scaly; scales more or less contiguous, the colour variable on one leaf. Petioles up to 9 mm long. Inflorescence of 6(-10) flowers in dense, terminal clusters. Pedicels very short, 1-2(-5) mm long, scaly. Calyx leafy, lobes up to 6 mm long, broadly oblong, sparsely scaly at base and along median line, or glabrous, margin ciliate. Corolla occasionally white, usually pink, mauve, indigo-blue or intense purple, widely funnel-shaped with spreading lobes, 10-20 mm long, tube 4-9 mm long, throat usually filled with long, curled white hairs, corolla often externally pubescent also. Stamens 10, exserted, 9-20 mm long, filaments red, pubescent near base. Ovary conical, ribbed, scaly, sometimes with a tuft of hairs at the insertion of the style, which persists as the fruit develops. Style pinkish red, 14-20 mm long, more or less pubescent in the lower half. Capsule ovoid, 4-6 mm long, scaly, and with persistent calyx.

Distribution. China (north-west & west Yunnan, south-west Sichuan), north-east Upper Burma.

Habitat. Alpine pasture, forest margins, 3350-4270 m.

Rhododendron russatum
VALERIE PRICE

RHODODENDRON HIRSUTUM

Subgenus *Rhododendron*
Section *Rhododendron* subsection *Rhododendron*

Rhododendron hirsutum is a very close relative of *R.ferrugineum* with which it vicariates, hybridizes and shares the vernacular name of 'Alpenrose'. Unlike the latter species, which prefers acid soil and shady locations, *R. hirsutum* grows in dry, stony situations in calcareous areas of the alps of southern central Europe, but not being as fussy as its relative, it will tolerate somewhat acid soils if the need arises. *Rhododendron hirsutum* does not form extensive pure stands as does *R. ferrugineum* . It further differs from that species in that it is, as its name implies, a bristly plant; the under surface of the leaves is less scaly and both calyx-lobes and pedicels are usually longer than those of its ally.

Both species belong to subsection *Rhododendron* which comprises the whole of series *Ferrugineum sensu* Hutchinson, a total of three species, all European in distribution. *Rhododendron myrtifolium* is the third member, which includes *R. kotschyi* in synonymy. The subsection is related to subsections *Lapponica* and *Rhodorastra*, but is distinct.

Rhododendron hirsutum is a good evergreen plant for the rock-garden, producing masses of flowers which are generally paler than those of *R. ferrugineum*. There are several forms in cultivation, including a double-flowered 'Flore Pleno', and a white-flowered variant (var. *albiflorum*) which gave rise in cultivation to a plant with incised leaves known as 'Albiflorum Laciniatum'. Var. *latifolium*, with broad, indeed almost round, leaves is not in cultivation, according to Davidian (1982). *Rhododendron* × *intermedium* is a hybrid between *R. hirsutum* and *R. ferrugineum*, and occurs in the wild where the species' distributions overlap. It, too, is available commercially. Propagation of *R. hirsutum* can be achieved by layering or by cuttings which will usually root readily. Alternatively plants can be raised from seed. *Rhododendron hirsutum*, a hardy, late-flowering species, was the first rhododendron to be grown in cultivation, and it was introduced by John Tradescant Jnr as early as 1656. Linnaeus published a description of the species in 1753 in his *Species Plantarum*. The accompanying plate (tab. 1853) was painted in 1815 or 1816 from material supplied by John Walker of Arnos Grove. The name of the artist is not recorded, but the engraver was Weddell.

Rhododendron hirsutum Linn., Sp. Plant. 392 (1753).

Description. A somewhat scaly, evergreen shrub up to 1 m in height; young growth pubescent and laxly scaly and with some loriform hairs. Leaves coriaceous, narrowly obovate, obovate or orbicular, (8-)13-30 mm long, (4-)7-14 mm wide, with rounded or tapered base and obtuse, mucronate apex, margins loriform-ciliate, dark green, glossy and glabrous above, paler green and sparsely scaly beneath, scales golden brown. Petioles 2-4 mm long, scaly and bristly. Inflorescence terminal, racemose, 4-12-flowered; rachis up to 1(-3) cm long, scaly and minutely pubescent. Pedicels 1-1.2(-2.5) cm long, sparsely scaly and minutely pubescent. Calyx 5-lobed; lobes narrowly triangular, lanceolate or oblong, 2-5 mm long, externally scaly, loriform-ciliate. Corolla pale to deep pink, or red, occasionally white, tubular-campanulate, 1.2-1.8(-2) cm long, sparsely scaly and pubescent externally, hairy within; tube 6-10 mm long; lobes spreading, ciliate. Stamens 10, unequal, usually as long as corolla-tube; filaments pubescent in lower part. Ovary scaly. Style approximately 2-3 mm long, sparsely pubescent at base. Capsule oblong or globose, (3-)5-6(-7) mm long, scaly with persistent calyx.

Distribution. Mountains of France, Germany, Switzerland, Austria, Italy, and Yugoslavia.

Habitat. Open woodland and scrub, on screes and slopes, 360-1850 m.

Rhododendron hirsutum
ENGRAVING BY WEDDELL

RHODODENDRON DAURICUM

Subgenus *Rhododendron*
Section *Rhododendron* subsection *Rhodorastra*

Rhododendron dauricum is a deciduous or semi-evergreen member of the very small subsection *Rhodorastra*, which includes only one other species, *R. mucronulatum*. In the Russian literature two additional species have been described (*R. ledebourii* and *R. sichotense*) but until more material is freely available for study, whether or not these are merely forms of the widely distributed and variable *R. dauricum* will remain unclear. Sleumer (1949) treated this subsection as a distinct subgenus, but Cullen (1980) relates it to subsections *Rhododendron* and *Lapponica* and gives it equivalent status in section *Rhododendron*.

The evergreen variety (var. *sempervirens*) of this species has been figured twice for *Curtis's Botanical Magazine*. The original account (in 1817) used the spelling 'dahuricum', but by the time Cowan prepared the second account in 1921, to accompany tab. 8930, the spelling had been altered to *dauricum*. The material for the original plate engraved by Weddell (tab. 1888) was supplied by Messrs Whitley, Brame and Milne of Fulham. Despite its different aspect, Cowan considered the plant to be identical to that portrayed in tab. 8930, which was painted from a plant growing in the Royal Botanic Garden at Edinburgh. In his 1980 revision, Cullen cites the latter plate under *R. dauricum*, thus confirming the identification. Var. *sempervirens* is said to have been introduced from Russia by Thomas Bell in 1798, although it was known in English gardens as early as 1780. More recently, in 1967, Robert de Belder re-introduced the variety by sending plants to Britain which had been raised from seed. The seed had been received from Russia under the name *R. ledebourii*. An attractive plant, *R. dauricum* bears large, saucer-shaped, rosy purple or pink flowers from December to February, strongly reminiscent of those of subsection *Saluenensia*, but is a much larger shrub, capable of attaining 3 m in height. Var. *sempervirens* differs in several ways from the type, as well as in being evergreen; the branching habit is more compact and twiggy; the variety flowers later than the species, in March and April and its bud-scales are more persistent than those of the species. Cowan mentions that in var. *sempervirens*, the scales of the leaf under-surface are less densely distributed and the flowers are often in pairs. In contrast the leaves of the type are densely scaly beneath and the flowers usually solitary.

A dwarf form of *R. dauricum*, with neat, glossy leaves and rosy purple flowers in March and April has been recently introduced from Japan, where it is used as a subject for Bonsai and container planting. Also in Japan, a clone with red-striped white flowers has been developed, but these plants do not come true from seed. *R. dauricum* has been used in hybridization. The most successful and widely grown hybrid is *R.* × *praecox* (*R. dauricum* × *R. ciliatum*) in which the flowers are precocious. A semi-evergreen clone called 'Midwinter' received an Award of Merit at an RHS show in 1963 and a First Class Certificate in 1969 when exhibited by the Crown Estate Commissioners, Windsor.

Rhododendron dauricum Linn., Sp. Plant. 392 (1753).

Description. A straggling shrub 0.5-1.5 m in height, with scaly and pubescent young growth. Leaves deciduous, or somewhat more persistent, coriaceous, oblong-elliptic, 1-3.6 cm long, 5-20 mm wide with obtuse base, obtuse or retuse, mucronate apex and crenulate margin. Upper surface glossy, dark green, glabrous except for minutely puberulous midrib; lower surface pale green and densely scaly. Petioles up to 5 mm long, more or less scaly. Inflorescence 1-3-flowered, terminal and axillary, often 2 or 3 inflorescences clustered at ends of branches. Pedicels short, almost obsolete, up to 1 mm long, scaly and minutely puberulous. Calyx rim-like or saucer-shaped, densely scaly. Corolla pale to deep rosy purple or pink or white, open funnel-shaped 1.4-2.1 cm long, 2-3.5 cm in diameter; tube darker in colour than lobes, short, wide, 5-11 mm long, externally hairy near base and very sparsely scaly, glabrous within; lobes spreading. Stamens 10, unequal, more or less exserted; filaments pink, pubescent in lower part. Ovary conoid, scaly. Style slender, about 2.5 cm long, exceeding corolla, glabrous. Stigma a disk. Capsule ovoid, scaly.

Distribution. USSR, (common and widely distributed in many different habitats), Mongolia and adjacent northern China and Japan.

Habitat. Most usually in open scrub, above 180 m.

Rhododendron dauricum
ENGRAVING BY WEDDELL

RHODODENDRON CALOSTROTUM subsp. CALOSTROTUM

Subgenus *Rhododendron*

Section *Rhododendron* subsection *Saluenensia*

This hardy member of subsection *Saluenensia* is one which is recommended by both Peter Cox and Desmond Clarke. That it is a highly desirable subject for the small garden cannot be gainsaid, as *R. calostrotum* for all its small stature — it barely attains 1 m in height in its largest forms — produces so many large, rosy purple flowers, (often measuring as much as 4 cm in diameter) that they completely conceal the plant which bears them. Kingdon Ward referred to the plant as 'the pigmy rhododendron with the giant flowers'. It has to be conceded that several species in this subsection are superficially very much alike, but from the grower's point of view, there can be no doubt that *R. calostrotum* is the best of them all, and its popularity is supported by the three awards which the species has won at RHS shows over the years. The earliest award was an Award of Merit, received by Lt. Col. L.C.R. Messel of Nymans, Sussex in 1935 for a plant with deep rosy mauve to magenta flowers (Forrest 27065). This was followed by an AGM in 1969. Then in 1971, the Cox's of Glendoick exhibited their clone 'Gigha' which was awarded a First Class Certificate.

Rhododendron calostrotum is an ideal plant for the rock garden. It produces flowers in both spring and autumn and fertile seed from which new stock can easily be raised. Propagation from cuttings is also feasible. An essential consideration when growing this species is that the roots are never allowed to dry out. Several colour forms are available commercially.

Farrer, Cox and Kingdon Ward all collected seed of *R. calostrotum* in 1919. The first two collectors found the species growing on Hpawshi Bum in Upper Burma, and Kingdon Ward's seed was retrieved from material growing on a granite ridge on the isolated peak of Imaw Bum (4000 m) about 32 km north west of Hpawshi Bum. Forrest subsequently collected seeds from other localities in the area. The plants raised from Kingdon Ward's seed were prostrate in habit, producing mats of thin, wiry stems and were, for a time, considered to be a different species called *R. keleticum* (now classed as a subspecies of *R calostrotum*; Kingdon Ward retains the credit for the introduction). Stapf gives a very full account of the species under tab. 9001, which Lilian Snelling painted from a plant raised by Mr E.J.P. Magor, Lamellen, Cornwall from Farrer 1045. The seed had been obtained from Sir George Holford and the plant first flowered at Kew in April 1923. Stapf records the enthusiasm of both Farrer and Kingdon Ward for this plant, and tells how Farrer found three species of *Rhododendron* growing near the summit of Hpawshi Bum at an altitude of over 3700 m, one of which was the species under discussion. Farrer (1919), noted that the trio grew 'all over the open brows and rocks', and that *R. calostrotum* 'covers the barest open braes and tops of moorland in a close flat carpet of dark foliage from which, on pedicels (in pairs) of an inch or so, rise large round blossoms of a rich warm magenta-rose'. The species is found in a variety of other habitats too, ranging from open stony ground to swamps, usually on acid soil. The var. *calciphilum*, as its name suggests, is a form of the species found growing on limestone. It was introduced by Kingdon Ward in 1926 (no. 6984) and Yü in 1938.

The taxonomy of subsection *Saluenensia* is complicated, and treated very differently by the various authors who have attempted to revise the group. Hutchinson (1930) listed 11 species and Davidian more recently recognized eight. In 1980, Cullen published his revision of subgenus *Rhododendron* in which he discussed this subsection at length. Suffice it to say here that he has reduced the number of species to two, one of which, *R saluenense*, includes two subspecies, while the other, *R. calostrotum*, comprises four subspecies. Cullen considers the subsection as a whole to be related most closely to subsection *Fragariflora* and more distantly to subsection *Uniflora*. This arrangement also leaves much to be desired, and is, as Peter Cox has commented, particularly unpopular with gardeners.

Rhododendron calostrotum Balf. fil. & K. Ward in Notes Roy. Bot. Gard. Edinburgh 13: 85 (1920) subsp. **calostrotum**.

Description. An erect, prostrate or mat-forming, densely scaly shrub from a few centimetres to 1.5 m in height. Leaves coriaceous, suborbicular to oblong-ovate or obovate or elliptic, 11-35 mm long, (2-)4-20 mm wide, with rounded base and obtuse, mucronate apex; glaucous or not above, with scales persisting when dried out, densely brown-scaly and often glaucous beneath, the scales overlapping and flaky, some stipitate and cup-shaped; margin loriform-ciliate. Petiole 2-5 mm long, scaly. Inflorescence terminal, 1-3(-5)-flowered. Pedicels 9-16(-27) mm long, scaly and with some stipitate, cup-shaped scales. Calyx leafy, deeply 5-lobed; lobes stained red, oblong-ovate, 3-8 mm long, 3-4 mm wide, externally variably scaly and minutely pubescent, puberulent within; margins loriform-ciliate. Corolla various shades of pink to rosy purple, magenta or deep purple (no white form is known), sometimes the upper lobes speckled with darker crimson or purple, widely funnel-shaped, 5-lobed, 2.5-3.8 cm in diameter, 1.3-2.8 cm long; tube 0.7-1.2(-1.4) cm long, externally hairy, occasionally also scaly. Stamens 10, exserted from throat, but not exceeding corolla, subequal, 6-17 mm long; filaments villous at base. Ovary conoid, scaly. Style red, slender, exceeding stamens, glabrous or with a few basal hairs. Stigma capitate, lobulate. Capsule 4-9 mm long, densely scaly, concealed by the persistent calyx.

Distribution. North-east and east Upper Burma, Assam, China (north-west Yunnan and south-east Xizang).

Habitat. Exposed positions, usually on acid soils, 3050-4880 m.

Rhododendron calostrotum subsp. *calostrotum*
LILIAN SNELLING

RHODODENDRON FRAGARIFLORUM

Subgenus *Rhododendron*
Section *Rhododendron* subsection *Fragariflora*

This dwarf species is one of the smallest, rarely exceeding 20 or 22 cm in height. It does not fit easily into any of the existing subseries or subsections and seems to be intermediate between subsections *Saluenensia* and *Campylogyna*. Cullen has therefore placed it in its own monotypic subsection *Fragariflora*. Davidian, however, includes it in series *Lapponicum* on the basis of its entire scales. *Rhododendron fragariflorum* appears to be most closely allied to *R. setosum* which is also aberrant and which Cullen puts into subsection *Lapponica* and Davidian retains in series *Lapponicum*. Whatever its affinities, the species is distinctive and easily recognizable with a neat, tufted habit and usually paired, strawberry-pink to purple, funnelform-rotate flowers. Although hardy, this is not an easy species to grow successfully and is apparently rare in cultivation, probably because it is difficult to propagate. Nevertheless it is well worth trying in the rock-garden, where it is most likely to succeed. Once established it will develop a neat, low tussock charmingly studded with blooms in May and June. If it continues to flourish, the plant will in time become mat-forming, tumbling over rocks and open ground as it does in its natural habitat, but it must be admitted that this stage of development is rarely achieved in gardens.

Rhododendron fragariflorum was discovered and introduced by Frank Kingdon Ward in 1924. It was growing among other dwarfs on the Temo La and Nyima La, north of the Tsangpo bend in south Xizang, at altitudes of between 4200 and 4500 m. Kingdon Ward collected it twice from this region and again in Assam, near the Bhutan frontier. Subsequently Ludlow and Sherriff found the plant in south and south-east Xizang and in Bhutan, and it was re-introduced by these collectors in 1947. The accompanying plate was prepared by Mary Mendum from herbarium material of Kingdon Ward 5734, collected in China (south Xizang, Temo La) in 1924.

Rhododendron fragariflorum K. Ward in Gard. Chron. 86: 504 (1929).

Description. A dwarf, evergreen, tussock and mat-forming, much-branched shrublet up to 23(-40) cm in height with short, stout branchlets; young growth scaly and puberulent. Leaves aromatic, oblong-elliptic, 10-17 mm long, 5-9 mm wide, with more or less rounded base, obtuse or rounded apex and crenulate margin, loriform-ciliate (when young), upper surface dark glossy green, rugose and persistently scaly, with puberulent midrib, lower surface pale green, reticulate, with golden brown, distant, vesicular scales. Petioles 1-2 mm long. Inflorescence terminal, 2-3(-6)-flowered. Pedicels crimson, 7-10 mm long, scaly and densely pubescent, with stipitate scales. Calyx reddish to crimson, divided to the base into 5 lobes, each lobe oblong with rounded apex, 5-7 mm long, sparsely scaly externally and sometimes puberulent, the margins scaly and hairy. Corolla strawberry-pink to purple, 13-18 mm long; tube 5-7 mm long, usually externally glabrous, pubescent within. Stamens 10, unequal, 8-12 mm long, exserted; filaments pubescent towards base. Ovary scaly. Style often red, exceeding stamens, glabrous. Capsule about 7 mm long, scaly.

Distribution. Bhutan, China (south-east Xizang).

Habitat. Open hillsides, swampy grassland, 3650-4500 m.

Rhododendron fragariflorum
MARY MENDUM

RHODODENDRON UNIFLORUM var. IMPERATOR

Subgenus *Rhododendron*
Section *Rhododendron* subsection *Uniflora*

Kingdon Ward was the first to find and introduce this apparently rare, mat-forming plant in 1926: he dubbed it 'Purple Emperor'. It was growing in northern Burma near the Diphuk La, about 39 km due north of Fort Hertz, at the source of the Seinghku river at an altitude of 3000 to 3600 m. He found only one plant in flower, so he returned in October of the same year and was able to send good seed (no. 6884) back to England. The seed germinated well and many plants were raised, from which Hutchinson drew up and published his description, naming the plant '*imperator*' in deference to Kingdon Ward's nickname for it.

Rhododendron uniflorum var. *imperator* certainly merits its name, the flowers are indeed 'imperial purple' - a glorious colour; they are also numerous and large considering the size of the plant that bears them.

Hutchinson placed the species in the series *Lepidotum*, from which it has since been transferred to subsection *Uniflora* and reduced to the rank of variety by Cullen (1978). The shape of the leaf-tips and the geographical distribution distinguish var. *imperator* from var. *uniflorum* which has rounded leaf-tips and comes from south-east Xizang; var. *imperator* has acute leaf-tips and grows at lower altitudes in north-east Burma. Both varieties are known only from the type collection and cultivated material. Subsection *Uniflora* is a small group comprising four species of which three, (*R. uniflorum*, *R. pumilum*, *R. pemakoense*) are very similar. The fourth, *R. ludlowii* (see plate 42), is quite distinct and only included in the subsection provisionally. The subsection was established by Cowan and Davidian when they reviewed the series *Lepidotum*, *Boothii* and *Glaucum* in 1948. The relationships between subsection *Uniflora* and other subsections are complicated; subsections *Saluenensia* and *Fragariflora* appear to be the two most closely allied to it.

Tab. 514 (new series) was painted by Lilian Snelling from material grown at Edinburgh and supplied by Sir William Wright-Smith in May 1951.

In cultivation this species usually retains its mat-forming habit, but occasionally a taller form develops, for example, a plant from Windsor was shown in 1962, which was nearly 50 cm tall. Being very near the surface, the roots should have only a light covering of soil, so it is best to position the plant under an overhanging rock or in some similar situation where protection is afforded from both frost and drought. If this precaution is taken, the species proves tolerably hardy.

A particularly fine plant, exhibited by Lord Swaythling, Townhill Park, Southampton, received an Award of Merit from the RHS in 1934.

Rhododendron uniflorum K. Ward in Gard. Chron. 88: 299 (1930), var. **imperator** (K. Ward) Cullen in Notes Roy. Bot. Gard. Edinburgh 36: 113 (1978).
R. imperator K. Ward in Gard. Chron. 86: 299 (1930).

Description. A dwarf, evergreen, usually prostrate and mat-forming shrub with scaly, ascending young growth, up to 20 cm in height, occasionally more erect and attaining 50 cm in height; Leaves aromatic, elliptic or oblanceolate, 13-25 mm long, (3-)5-10 mm wide, with cuneate base and acute or rounded apex, margin revolute, upper surface very sparsely scaly, lower surface glaucous and sparsely scaly; scales small, golden, soon darkening. Petioles 2-4 mm long. Inflorescence 1-2(-3)-flowered, borne singly in the axils of upper leaves. Pedicels slender, stained red, 10-12(-15) mm long, extending to 25 mm in fruit, scaly. Calyx 1-2.5 mm long, 5-lobed, lobes ovate-oblong, obtuse, scaly. Corolla rosy purple, funnel-shaped, 21-25(-30) mm long; tube 12-14 mm long, externally densely short-hairy and sparsely scaly. Stamens 10, unequal, 19-25 mm long; filaments pubescent near base. Ovary ovoid, scaly. Style reddish, slender, about 25 mm long, exceeding stamens, glabrous, impressed. Stigma capitate, lobulate, 1.5 mm in diameter. Capsule about 9 mm long, scaly.

Distribution. north-east Burma.

Habitat. Bare cliff ledges, 3050-3350 m.

Rhododendron uniflorum var. *imperator*
LILIAN SNELLING

RHODODENDRON LUDLOWII

Subgenus *Rhododendron*
Section *Rhododendron* subsection *Uniflora*

In Stevenson's *The Species of Rhododendron*, *R. ludlowii* was placed in the series *Lepidotum*, but most of the species included in that series have a short, stout, deflexed style in sharp contrast to the long, slender style of *R. ludlowii*. The discrepancy did not go unnoticed and in 1948, Cowan and Davidian re-classified *R. ludlowii*, together with six other species, including them all in Sleumer's subsection *Uniflora*. In the recent Edinburgh revision of the genus *Rhododendron* (1980), Cullen has retained subsection *Uniflora*, to comprise only five taxa, i.e. 4 species (one species being split into two varieties). One of these species is *R. ludlowii*, which he admits is aberrant and only provisionally included in the subsection. The latter species is distinguished from the other four by its crenate leaves, large calyx and cupular, externally pubescent and scaly, yellow, spotted corolla. It has been used successfully by hybridists to produce dwarf yellow hybrids.

Mr Frank Ludlow and Major George Sherriff first collected this plant on the Lo La pass, Packakshiri, south-east Xizang, on the border between that province and Assam, in July 1936 (L & S 1895). It was growing at an altitude of 450 m in the alpine zone where the plants would normally be buried under about 2 m of snow in the winter. The same two collectors, in company with Sir George Taylor, again found the species in June 1938, this time at Tsari Sama, where it was locally common, growing on moss-covered rocky soil on open hillsides (L, S & T 5571) in association with Cassiope species and other dwarf rhododendrons. Seeds from the same locality, but a different gathering (L, S & T 6600) produced plants in cultivation which effected the first introduction of this beautiful but difficult species. Davidian (1982) noted that in cultivation the plant forms an erect, rounded shrub up to 30 cm high, instead of retaining its more usual prostrate, creeping habit. A full account of the cultivation of *R. ludlowii* is given in J.R. Sealy's text (illustrated by tab. 412 in *Curtis's Botanical Magazine*) to which reference should be made if the brief notes given here require amplification. In it Sealy has quoted from notes sent to Kew by Mr R.B. Cooke (Corbridge, Northumberland) with the material from which tab. 412 was painted by Margaret Stones in May 1957. The species is extremely slow-growing and is not entirely hardy, but is well worth pampering. It flowers at an early age and is very rewarding when well grown, producing numerous, single or paired, yellow, red-spotted flowers, large in relation to the size of the plant, at the branch-tips in May. Fertile seed is also produced. To grow *R. ludlowii* well is not easy, but the species responds to partial shade with ample organic matter in the soil and, to protect the rather feeble root system from heat, a few stones around the neck of the plant. Given these conditions, success may well be achieved.

Rhododendron ludlowii Cowan in Notes Roy. Bot. Gard. Edinburgh 19: 243 (1937).

Description. A dwarf, evergreen, prostrate and spreading shrub up to 30 cm in height, the young growth bearing more or less stalked scales. Leaves coriaceous, broadly obovate, oblong-obovate or obovate-elliptic, (9-)15-16 mm long, (5-)9-10 mm wide with cuneate or rounded base and obtuse, mucronate apex; margins crenate. Upper surface dark green with some pale scales; lower surface pale green or brownish, sparsely scaly with conspicuous venation. Petioles about 1 mm long. Inflorescence usually 1-flowered, occasionally 2-flowered, at branch tips. Pedicels 15-20(-25) mm long, stout, stained red, scaly. Calyx large, leafy, deeply divided into 5 oblong, rounded lobes 4-5.5 mm long, scaly and sparsely ciliate. Corolla yellow, often spotted with red, cup-shaped or widely funnel-campanulate, (1.5-)2-2.3(-3) cm long, 2.5-4 cm across; tube more or less 1.4 cm long, externally densely pubescent and scaly, densely pubescent at base within. Stamens 10, unequal, shorter than corolla, (7-)11-19 mm long; filaments pubescent near base. Ovary ovoid, densely scaly. Style slender, declinate, impressed, exceeding stamens, 20-23 mm long, glabrous. Stigma capitate, 1 mm in diameter. Capsule (4-)7-8 mm long, somewhat scaly with persistent calyx.

Distribution. *Rhododendron ludlowii* is known only from the collections detailed above, from China (south-east Xizang) and from cultivated material.

Habitat. Open hillsides, 4000-4300 m.

Rhododendron ludlowii
MARGARET STONES

RHODODENDRON CINNABARINUM subsp. CINNABARINUM

Subgenus *Rhododendron*
Section *Rhododendron* subsection *Cinnabarina*

In the Edinburgh revision (1980) subsection *Cinnabarina* contains subseries *Cinnabarinum* in its entirety and has altered little in the transference. It forms a well-defined Himalayan group, its eastern limit being the Mishmi Hills in Arunachal Pradesh and all the included species and varieties are said to be hexaploids. The subsection as a whole is allied to subsection *Tephropepla*. Cullen has split this species into three subspecies based on corolla and leaf characters, but despite the incredible colour range of the flowers and other easily recognizable differences, the majority of the forms of *R. cinnabarinum* are not considered to be worthy of taxonomic status. There are some very attractive plants which cannot be ignored, such as the one figured here, in tab. 4930, by Fitch.

Originally ranked at specific level, 'var. *blandfordiiflorum*' now included in subsp. *cinnabarinum*, produces waxy bi-coloured flowers which are particularly beautiful when backlit by the sun and are admirably set off by the neat, aromatic, glaucous foliage. This variety was introduced by Joseph Hooker in 1850 from eastern Nepal, where it was growing at altitudes of between 3000 and 4000 m. The flowers form the main attraction, as the plant itself is tall and spindly and the leaves few and far between. The bi-coloured form of subsp. *cinnabarinum* is still one of the most popular variations and it makes up for the above-mentioned shortcomings by flowering late, from May to July, thereby avoiding frost-damage. All the forms of *R. cinnabarinum* tend to be tall, the leaves sparse and the habit 'thin'. If you consider this mode of growth 'open and graceful' as one writer (I.F.LaCroix) put it, then when grown singly, few plants will displease; otherwise they are seen at their best planted either in a group of mixed forms, or in a shrubbery amongst leafier species. As well as the myriad forms, *R. cinnabarinum* has also produced many hybrids, some of which are considered to be an improvement on the parent. Two of these are the familiar 'Lady Roseberry' and 'Lady Chamberlain' grexes which are among the most frequently grown. Unfortunately the species is susceptible to powdery mildew which weakness it only too readily passes on to its progeny: this is a point to be borne in mind when buying stock. It should be noted that the plant is poisonous to cattle and goats, and the smoke from the burning wood causes inflammation of the eyes and swelling of the face in humans. A plant of *R. cinnabarinum* subsp. *cinnabarinum* shown as var. *blandfordiiflorum* received an Award of Merit from the RHS when exhibited in May 1945 by Lord Aberconway, Bodnant; a clone named 'Nepal' shown by Hydon Nurseries Ltd., Godalming and derived from the collection L, S & H 21283, also received an Award of Merit in 1977 — as *R. cinnabarinum*.

Rhododendron cinnabarinum Hook. fil., Rhodo. Sikkim Himal., t. 8 (1849). subsp. **cinnabarinum**.
R. roylei Hook. fil., Rhodo. Sikkim Himal., t. 7 (1849).
R. blandfordiiflorum Hook. in Bot. Mag. 82: t. 4930 (1856).
R. cinnabarinum var. *roylei* (Hook.) hort.
R. cinnabarinum var. *blandfordiiflorum* (Hook.) hort.

Description. A slender, spindly shrub up to 3(-7) m in height, with sparse, long, slender branches; young growth scaly, often glaucous or pruinose. Leaves usually evergreen, sometimes deciduous, more or less coriaceous, narrowly to broadly elliptic-lanceolate, 3-9(-11.5) cm long, 2.7-5.5 cm wide, with tapered to cordate base and rounded obtuse apex, upper surface glaucous green (especially the young leaves) with a metallic lustre, glabrous, lower surface glaucous green to reddish brown, scaly, the scales often unequal in size, midrib scaly above and beneath. Petioles 4-13 mm long, scaly. Inflorescence terminal, with 2-7(-9) nodding flowers. Pedicels 7-8 mm long, sometimes more, scaly. Calyx variable, rim-like or with 5 subequal or unequal lobes, one often much longer and narrower, scaly. Corolla extremely variable in colour, usually red (but in the bi-coloured form of the subspecies the tube is orange-red externally, yellow or greenish within, and the lobes are yellow or greenish), tubular-campanulate with copious nectar in basal pouches, 2.5-3.8(-5) cm long. Stamens 10, unequal, exserted from tube but not exceeding lobes; filaments pubescent towards base, rarely glabrous. Ovary conoid, tapering to style, ribbed, scaly, sometimes also pubescent at apex with a basal, lobulate, glabrous disk. Style slender, exceeding stamens, glabrous or pubescent or, rarely, scaly at base. Stigma clavate, lobulate. Capsule up to 1.3 cm long, scaly.

Distribution. East Nepal, India (west Bengal, Sikkim), Bhutan and as far east as the Tsangpo gorge in China (south-east Xizang).

Habitat. Scrub, open woodland, various types of forest, and on steep, rocky hillsides and cliffs, 2100-4000 m.

Rhododendron cinnabarinum subsp. *cinnabarinum*
WALTER HOOD FITCH

RHODODENDRON TEPHROPEPLUM

Subgenus *Rhododendron*
Section *Rhododendron* subsection *Tephropepla*

This variable, but always beautiful, species was discovered by Reginald Farrer in 1920 and introduced into cultivation by George Forrest in 1921. Farrer's plant was abundant at an altitude of 3200 m on the Chawchi and Maguchi passes between Nmai Hka and the Salween in north-east Upper Burma, but Forrest gathered his material from a more northern location along the same divide, in east Xizang. The species was introduced many times subsequently and became a great favourite with Lionel de Rothschild, as with many other enthusiasts. That it did so is not surprising, for the species is not only very free-flowering but, apart from suffering occasional frost damage to the flowers and buds, is usually hardy. It needs protection only from hot sun, so is best grown on the edges of woodland where it is afforded a certain amount of shelter, and where the full beauty of the masses of bright pink flowers shows to advantage against a green backdrop.

Being a variable species, numerous forms have been introduced by the various collectors and others have been developed by nurserymen, with the result that a good selection is available commercially. This popular species can be tall or short, small or large, and the flower colour ranges from pure white to deep magenta-pink. The leaves, too, vary considerably in size. Planted in a mixed group, these forms put on a breathtaking display during April and May.

Rhododendron tephropeplum has received three awards from the RHS, all Awards of Merit. The first was in 1929, for a plant from Bodnant, exhibited by Lady Aberconway and the Hon. H.D. McLaren; the second award was given in 1935 to a plant exhibited as *R. deleiense* by Lord Swaythling of Townhill Park, Southampton (this name is now listed as a synonym of *R. tephropeplum*). A form from Kingdon Ward 20844, named 'Butcher Wood', and shown by Major A.E. Hardy of Sandling Park, Kent, claimed the third award in 1975.

In the Bayley Balfour system of classification, *R. tephropeplum* was included in the *Boothii* series, with which it showed little affinity, but where it was considered to be best placed, having no obvious relationships with any other series. In the Edinburgh revision (1980), Cullen created a subsection *Tephropepla* to accommodate this species and four others, all of which have characteristics connecting them to subsections *Cinnabarina* and *Virgata*.

Lilian Snelling's plate (tab.9343) was painted from material supplied by Mr J.C. Williams of Caerhays Castle, Cornwall.

An interesting note appears in Hutchinson's account of the plant depicted in tab. 9343, where he states that on examining the scales of this species, Miss E.M. Wakefield discovered that the presence of the mycelium of a small mould/fungus, growing on the gummy exudation at the centre of the scales, was responsible for their dark coloration. He added that the mould was absent from the leaves of plants grown in Britain.

Rhododendron tephropeplum Balf. fil. & Farrer in Notes Roy. Bot. Gard. Edinburgh 13: 302 (1922).
R. spodopeplum Balf. fil. & Farrer, *op. cit.* 299.
R. deleiense Hutch. & K. Ward in Notes Roy. Bot. Gard. Edinburgh 16: 172 (1931).

Description. An evergreen, scaly shrub, small and compact to erect and leggy, up to 1.5(-2.4) m in height with brownish, scaling bark and slender branches. Leaves coriaceous, narrowly oblanceolate to narrowly elliptic, (3-)5-7.5(-13) cm long, (1-)1.6-3(-4) cm wide, with cuneate base and rounded, mucronate apex, margins somewhat recurved, upper surface dark green, sparsely scaly, soon glabrous, lower surface scaly and grey-papillose, speckled brown, finally evenly brownish grey as scales darken; scales unequal. Petioles 5-7 mm long, scaly. Inflorescence umbellate, 3-9-flowered; rachis 1-4 mm long. Pedicels (1.1-)1.6-1.8(-3) cm long, densely scaly. Calyx large and leafy, 5-lobed; lobes orbicular or oblong, rounded at the apex, 5-7(-8) mm long, scaly at base and sparsely loriform-ciliate. Corolla white, pink, magenta or red, campanulate, (1.7-)2-2.4(-3.2) cm long overall; tube 1.1-1.8 cm long, externally scaly, otherwise glabrous; lobes spreading to recurved, emarginate. Stamens 10, often forming a tight ring in the mouth of the tube, 1-2.3 cm long; filaments minutely pubescent near the base. Ovary conoid or ovoid, ribbed, scaly. Style slender, straight, exceeding stamens, lower half scaly. Stigma crimson, large. Capsule ovoid-cylindrical, (5-)7-10 mm long, with persistent calyx.

Distribution. North-east India (Arunachal Pradesh), north-east Burma and China (north-west Yunnan, south-east Xizang).

Habitat. Cliffs, rocky slopes and screes, in thickets and in alpine meadows, 2450-4300 m.

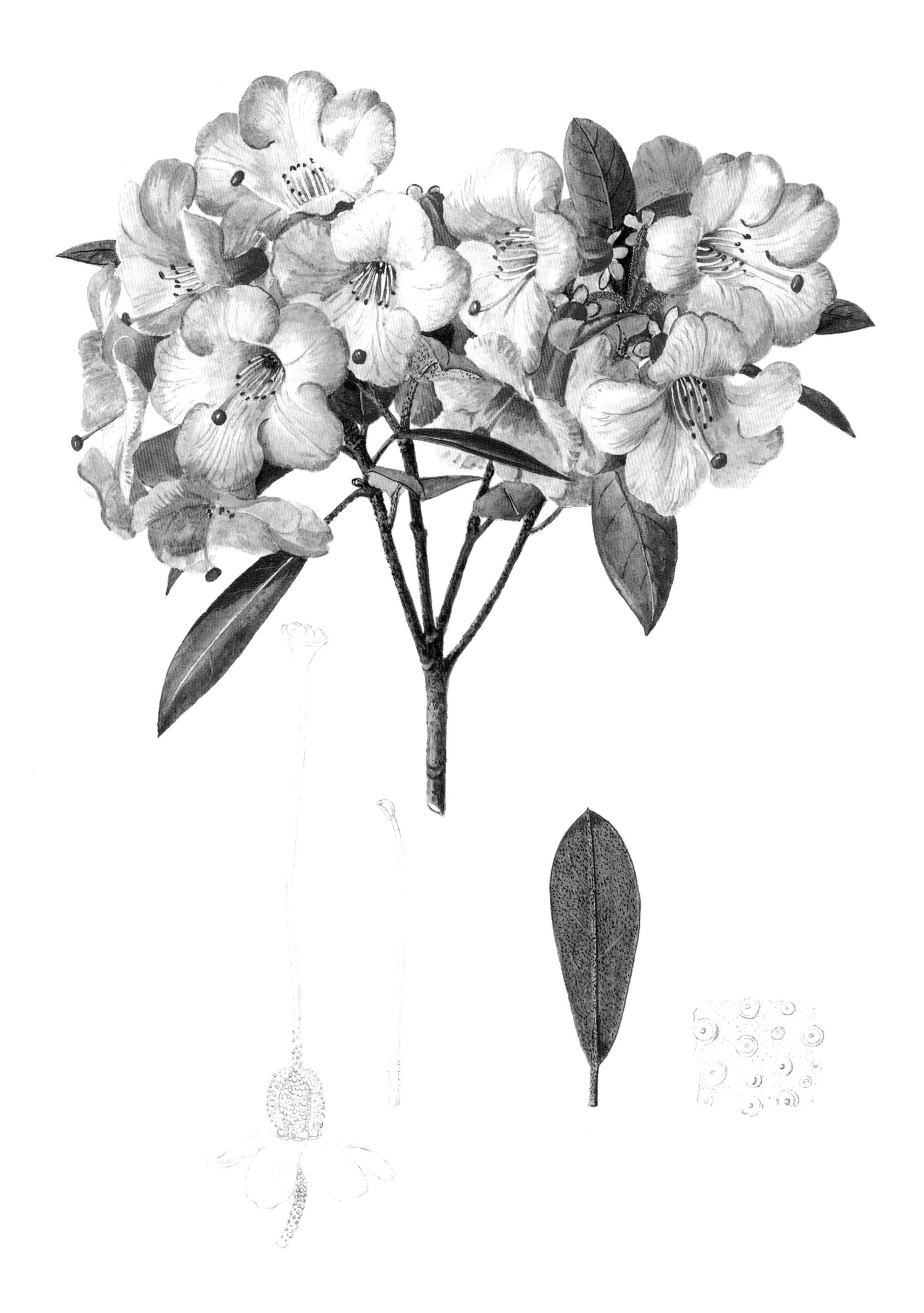

Rhododendron tephropeplum
LILIAN SNELLING

RHODODENDRON VIRGATUM subsp. VIRGATUM

Subgenus *Rhododendron*
Section *Rhododendron* subsection *Virgata*

The material from which Fitch prepared tab. 5060 was taken from a plant grown from seeds collected by Booth in Bhutan. The seed was sent to Nuttall, who in turn passed it to Mr Lowe of Clapton Nursery. The plants which were raised there, flowered there in 1858, the year that the plate was painted.

Joseph Hooker considered this variable species to be synonymous with *R. triflorum*, but his error was noted in the text accompanying tab. 5060 and therefore caused very little confusion. It was Hooker who first discovered the species while travelling in the Sikkim Himalaya in May 1849, and introduced it into cultivation in the following year. He found it growing abundantly on the margins of pine forests in ravines in the Lachen Valley at about 2000-3000 m, where it was usually confined to dry areas. The species ranges from Nepal to Yunnan in south-west China, with subsp. *oleifolium* replacing subsp. *virgatum* in the east. Despite several re-introductions, the species is still not common in cultivation; this is possibly partly due to its wide altitudinal range, as a consequence of which, *R. virgatum* is variable in hardiness, some forms being particularly tender and difficult to please. It is variable too in height, leaf shape and size and flower size, the smaller forms being especially suitable for the rock-garden, where they appreciate the open, dry, sunny positions. The larger forms tend to sprawl and are best situated on a bank where they can 'tumble' without smothering smaller species growing round their bases. This species has the merit of being easy to propagate from cuttings and is free-flowering, with white or more usually pink, rose-coloured or mauve corollas, produced from March to May.

Rhododendron virgatum received an Award of Merit from the RHS when shown in 1973 by Major A.E. Hardy of Sandling Park, Kent.

In New Zealand, the species has been successfully crossed with the Australian *R. lochae*, which perhaps suggests a closer affinity with the Malesian species than has yet been considered likely. It certainly offers scope for more experimentation and research. At present *R. virgatum* forms the monotypic subsection *Virgata*, hitherto classed as series *Virgatum* Hutch. The subsection is probably related to subsections *Cinnabarina* and *Tephropepla*.

Rhododendron virgatum Hook. fil., Rhodo. Sikkim Himal., t. 26 (1849).

Description. An erect or sprawling, evergreen shrub up to 1.5(-2.4) m in height with scaly young growth. Leaves scattered along the branches, coriaceous, up to 5(-8) cm long, 0.5-2 cm wide, oblong-elliptic or narrowly so, with tapered or more or less rounded base and mucronate apex; upper surface laxly scaly, especially along the midrib and at the base; lower surface densely scaly and more or less glaucous-papillose, the scales unequal, flaky; margins revolute. Petioles 3-7 mm long. Inflorescence of usually solitary or (less often) paired, fragrant flowers in each axil of the upper leaves and sometimes for the top several centimetres of the shoots. Pedicels 2-8 mm long, concealed by the persistent floral bud-scales. Calyx 5-lobed, the lobes rounded, occasionally ciliolate, 2-3 mm long, externally scaly. Corolla funnel-form, 1.5-3.7(-3.9) cm long; tube 0.8-2 cm long, externally pubescent and sparsely scaly, the scales extending onto the 5 lobes. Stamens 10, unequal, somewhat exserted from, or often forming a ring in the mouth of the tube, 9-38 mm long; filaments pubescent at base. Ovary ovoid, 2-3(-4) mm long, densely scaly. Style slender, widening upwards, exceeding stamens, scaly and/or pubescent near the base. Stigma wider than the style. Capsule more or less cylindrical, (5-)9-12.5 mm long, densely scaly, with persistent calyx. Seeds with a tail at each end.

Distribution. Subsp. *virgatum* is found in Nepal, India (Arunachal Pradesh, Sikkim) Bhutan and China (south-east Xizang). Subsp. *oleifolium*, which used to be known as *R. oleifolium*, is known only from south-east Xizang and the western parts of Yunnan. It is further distinguished from subsp. *virgatum* by its smaller, white or pink corolla (1.5-2.5 cm. long; tube 0.8-1.5 cm. long).

Habitat. Forest margins, stony slopes and in scrub, 2500-3800 m.

Rhododendron virgatum
WALTER HOOD FITCH

RHODODENDRON MICRANTHUM

Subgenus *Rhododendron*
Section *Rhododendron* subsection *Micrantha*

The account of this hardy, late-flowering species in volume 134 (tab. 8198) of *Curtis's Botanical Magazine* was written by two authors: T.F. Chipp wrote the text and W. Watson contributed the notes on cultivation.

The initial collection of *R. micranthum* was by the French missionary, d'Incarville, in the mountains north of Beijing (Peking) and it was from the same area that a later collection formed the basis of Turczaninow's formal description in 1837. The first introduction into cultivation was accomplished by E.H. Wilson through Veitch in 1901, after he found *R. micranthum* growing in west Hupei. He re-introduced it in 1907, 1908 and 1910, but he collected it only in China, where the species is widely distributed in scrub or forests, on grassy slopes, cliffs, ridges and in dry gorges at high elevations.

Rhododendron micranthum is not a showy plant, and in fact is more reminiscent of a large *Ledum* than the popular conception of a *Rhododendron*. It is a free-flowering species which has a quiet attraction of its own when covered with its numerous many-flowered terminal racemes (not 'lateral corymbs' as stated by Turczaninow). It first flowered in cultivation in May 1904 at Veitch's nursery, Coombe Wood; the flowering period usually extends into June and July.

Tab. 8198 was painted by Matilda Smith from a plant which was raised at Caerhays Castle, Gorran, Cornwall, and was supplied by J. C. Williams. It was presented to Kew through Veitch.

Rhododendron micranthum forms Cullen's monotypic subsection *Micrantha* which does not have any close allies, only more or less tenuous links with subsection *Lapponica*. Its seeds are very distinct, with conspicuous wings, which indicate connections with subsections *Maddenia* and/or *Boothia*.

Rhododendron micranthum Turcz. in Bull. Soc. Nat. Mosc. 7: 155 (1837), fig. lk, p. 15.
R. rosthornii Diels in Bot. Jahrb. Syst. 29: 509 (1900).
R. pritzelianum Diels, *op. cit.*: 510 (1900).

Description. A straggly shrub up to 2(-5) m in height, with scaly, minutely pubescent young growth and long, slender branches. Leaves coriaceous, persisting for 1 year, oblanceolate-elliptic or narrowly so, (16-)34(-59) mm long, (4-)9-25 mm wide with gradually tapering, cuneate base and acute, mucronate apex. Upper surface of midrib and some veins, minutely pubescent and sparsely scaly, lower surface more or less densely scaly, scales light brown, usually contiguous or overlapping. Inflorescence a terminal or occasionally (in the uppermost leaf-axils) axillary, rounded 20(-28)-flowered raceme with conspicuous pubescent rachis 1-2.6 cm long. Pedicels slender, (7-)10-20 mm long, puberulent and sparsely scaly. Calyx 5-lobed, lobes triangular, (0.5-)1-2 mm long, acute, externally scaly with loriform-ciliate margins. Corolla creamy white, funnel-campanulate, deeply divided into 5 lobes, 4-8 mm long; tube 1-3 mm long, externally densely scaly, glabrous within. Stamens 10, unequal, 4-8 mm long, exceeding corolla and style; filaments glabrous. Ovary conoid, scaly. Style slender, glabrous or with a few hairs near the base. Capsule oblong-cylindric, 4-6(-8) mm long, scaly. Seeds conspicuously winged.

Distribution. Widely distributed over north & central China from west Sichuan to the Beijing (Peking) district, Manchuria, Korea.

Habitat. Montane scrub and thickets, 1600-2600(-3000) m.

Rhododendron micranthum
MATILDA SMITH

RHODODENDRON LEUCASPIS

Subgenus *Rhododendron*
Section *Rhododendron* subsection *Boothia*

Rhododendron leucaspis is a charming hardy species which was introduced into cultivation by Frank Kingdon Ward. It received an Award of Merit from the RHS when a plant raised from KW 6273 was exhibited in 1929 by Mr L. de Rothschild, Exbury. The species also received a First Class Certificate in 1944 when another Kingdon Ward plant, this time derived from KW 7171 was shown by Mr E. de Rothschild.

When discovered in November 1924, Kingdon Ward's no. 6273 was growing with other rhododendrons and bamboos on steep grassy slopes and on cliffs in the Tsangpo Gorge in Xizang. The collection was made at Musi La, below Gyala at an altitude of 3050 m. Plants raised from seeds of this gathering flowered in the spring of 1928. Thus it appears that a short three years is the normal period taken by the species to reach its flowering stage, a characteristic which endears it to growers! A plant raised at Kew from the same KW number, flowered in late March 1940 when it was painted by Stella Ross-Craig for *Curtis's Botanical Magazine* (tab.9665). Two years later, Kingdon Ward again collected the species, this time in the Di Chu valley on the Xizang-Burma frontier, about 320 km south-east of the Tsangpo Gorge, where it was growing in damp, heavily shaded conditions. This was KW 7171.

Rhododendron leucaspis is a well-known species, and is easily identifiable by the numerous flat, white, broad-petalled flowers cradled in red-tinged bud-scales, for which it was named ('*leucaspis*', means 'white shield'), its conspicuous nutmeg-brown stamens and stubby, deflexed style. Unfortunately it blooms so early in the year that buds and flowers are frequently frosted. Bark-split can also be a problem, so some shelter should be provided if such damage is to be avoided. On occasion a plant reaches a height of a metre or so and in the wild the species is known to be epiphytic, although only rarely so. On the whole, however, *R. leucaspis* is best suited to the rock-garden where its normal habit is to sprawl, forming mats of low growth. It also makes an attractive container plant.

Subsection *Boothia*, as delimited by Cullen (1980) is a small group of seven species which includes most of the old series *Boothii*. *R. leucaspis* is somewhat aberrant, but is linked to the group by its vesicular scales and general similarity to *R. megeratum*. The vesicular scales resemble those of the allied subsection *Trichoclada*, while subsection *Boothia* as a whole is related to subsections *Edgeworthia*, *Maddenia*, *Camelliiflora* and *Glauca*.

Rhododendron leucaspis Tagg in Gard. Chron. 85: 128, 135, 308 (1929).

Description. A small shrub up to 1 m in height, but often mat-forming; young growth densely clothed with straight loriform setae. Leaves spaced along young branches, broadly elliptic, 3-4.5(-6) cm long, 1.8-2.2(-3) cm wide, with cuneate base and obtuse apex, upper surface and margins densely loriform-setose, lower surface glaucous with impressed vesicular scales. Petioles flat, ciliate. Inflorescence terminal, 1-2(-3)-flowered. Floral bud-scales variably persistent, reddish, shortly white-ciliate. Pedicels short, lepidote, sometimes pubescent also, and with a few loriform hairs. Corolla white, often stained pink, broadly campanulate to rotate, 2.5-3(-5) cm in diameter; tube very short, externally scaly; lobes 8-11 mm long, spreading to reflexed. Stamens 10, exserted, exceeding style; filaments about 15 mm long, pubescent in lower half to three-quarters; anthers conspicuous, dark brown, up to 4 mm long. Ovary conical, densely scaly. Style short, stout, deflexed, expanded beneath stigma. Stigma capitate, lobulate. Capsule up to 10 mm long, scaly.

Distribution. China (South Xizang, Tsangpo river gorge and other nearby gorges). Nowhere is it abundant, but both Kingdon Ward and Peter Cox suspected that it might be more common than is supposed, growing epiphytically.

Habitat. Cliff faces, grassy banks, 2440-3050 m.

Rhododendron leucaspis
STELLA ROSS-CRAIG

RHODODENDRON CAMELLIIFLORUM

Subgenus *Rhododendron*
Section *Rhododendron* subsection *Camelliiflora*

The affinities of this species are complex and it is therefore no surprise to find that it has been assigned to a monotypic subsection *Camelliiflora*. Closely related to subsection *Boothia*, it is distinguished from the latter subsection by its larger number of stamens (12-16) and ovary with five to ten locules. These characters also indicate some relationship to subsection *Maddenia* and Cullen comments on the fact that when not in flower, *R. camelliiflorum* is difficult to distinguish from *R. keysii* in subsection *Cinnabarina*. The same similarities are remarked upon in the original account accompanying tab. 4932 where there is additional comment that the corollas resemble those of yet another subsection — *Lepidota*!

Rhododendron camelliiflorum is an undistinguished species which on the whole has little to recommend it as a garden plant except as an example of the variability within the genus *Rhododendron*. The flowers are small, creamy white with a ring of contrasting rust-brown stamens, and are not particularly numerous; the plant also requires some form of protection in all but the mildest areas of Britain. The species was first discovered by J.D. Hooker, who found it growing at elevations of 2700-3600 m in both east Nepal and Sikkim in 1848. He included it in his book *Rhododendrons of the Sikkim Himalaya* and introduced it into cultivation at Kew in 1851, where tab. 4932 was painted by Fitch for *Curtis's Botanical Magazine*. It was also found by Griffith in the mountains of Bhutan, but has not been found outside these three regions. It may or may not be coincidental that plants of the species collected in Bhutan seem to produce consistently wider leaves than those from the other two areas. When growing epiphytically in restricted light, the species develops long, drooping branches well over a metre long, but in adequate light on forest margins, cliffs, etc, where it is usually terrestrial, the growth is much more compact.

Rhododendron camelliiflorum Hook. fil., Rhodo. Sikkim Himal. t. 28 (1849).
R. sparsiflorum Nutt. in Hooker's J. Bot. Kew Gard. Misc. 5: 363 (1853).
R. cooperi Balf. fil. in Notes Roy. Bot. Gard. Edinburgh 10: 91 (1917).

Description. An epiphytic or terrestrial, evergreen shrub up to 2 m in height, with branches up to 2 m long; young growth densely scaly. Leaves coriaceous, narrowly elliptic to oblong-elliptic, occasionally broadly oblong, 6-9(-13.5) cm long, (1.6-)2-3.5(-4.2) cm wide, with obtuse to rounded base and acute, mucronate apex, glossy, dark green above, sparsely scaly and soon glabrescent, matt and paler green or greenish brown beneath, densely covered with more or less circular, broadly rimmed scales and with a scattering of larger darker scales amongst them. Petioles stout, 4-10(-14) mm long, grooved above, densely scaly. Inflorescence axillary, 1-2(-3)-flowered. Floral bud-scales more or less persistent, ovate, ciliate and sparsely scaly. Pedicels 4-12 mm long, densely scaly. Calyx large, leafy, 5-lobed; lobes oblong with rounded apex, 5-8 mm long, scaly or not, margin scaly. Corolla fleshy, white, occasionally yellowish at base of tube, and often tinged with pink or deep pink, up to 4 cm across lobes; tube short and broad, 8-12 mm long, externally scaly, villous within; lobes spreading, broadly ovate, 1.4-1.5 cm long, 1.2-1.3 cm wide. Stamens 12-16, subequal, shortly exserted, usually exceeding style; filaments pubescent towards base; anthers conspicuous, rust to dark brown, about 3 mm long. Ovary conoid, 5-10-locular, densely scaly with lobed, glabrous basal disk. Style short, curved, up to 1.4 cm long, glabrous. Stigma clavate. Capsule obloid, (7-)10(-13) mm long, densely scaly with persistent style.

Distribution. Nepal, India (Sikkim) and Bhutan.

Habitat. Grows epiphytically or terrestrially in forests, in open situations on forest margins and on cliffs, 2750-3650 m.

Rhododendron camelliiflorum
WALTER HOOD FITCH

RHODODENDRON GLAUCOPHYLLUM var. GLAUCOPHYLLUM

Subgenus *Rhododendron*
Section *Rhododendron* subsection *Glauca*

Previously known as *R. glaucum* Hook. fil., this species has been transferred with several other species from the series of that name to subsection *Glauca* in the Edinburgh revision. It is now named *R. glaucophyllum* Rehder, and the name *R. glaucum* has been relegated to synonymy. Two varieties are recognized; var. *glaucophyllum* with campanulate corollas and sharply deflexed styles, and var. *tubiforme* with tubular-campanulate corollas and declinate styles — the latter may possibly be a natural hybrid. The subsection as a whole is related to subsection *Boothia*.

The plate in *Curtis's Botanical Magazine*, tab. 4721, painted by Fitch, shows var. *glaucophyllum*, a plant raised from seed collected by J.D. Hooker in the autumn of 1850 in the Sikkim Himalaya. This was its first introduction into cultivation and flowers were produced in Spring 1853, when the plant was about 30 cm in height, having been grown in a cool glasshouse at Kew. Those plants which had been grown outside had not at that time produced flowers. Subsequently, Ludlow and Sheriff introduced var. *tubiforme* from east Bhutan (L & S 52856) which Cullen speculates may be a natural hybrid, although this is by no means proven.

Rhododendron glaucophyllum is reasonably hardy and, as it does not flower until late April or May, often escapes the frosts. It is free-flowering, not especially showy, but where planted in groups, the resulting soft-pink cloud of flowers is most effective. A long-lived species, it has a spreading habit which in the wild produces large colonies; it will also cover large areas of the garden if left undisturbed for a number of years. Cox (1985) suggested that it is at its best along woodland edges, where it shows to advantage. A clone of the recently introduced white-flowered form from east Nepal, was given the cultivar name 'Len Beer', after one of its collectors (Lancaster and Morris were the others) when it was shown in 1977. Davidian gives this cultivar varietal status as var. *album*.

Rhododendron glaucophyllum Rehder in J. Arnold Arb. 26: 73 (1945) var. **glaucophyllum**.
R. glaucum Hook. fil., Rhodo. Sikkim Himal., t. 17 (1849).

Description. A lax, spreading, evergreen shrub up to 1.5 m in height with scaly branches. Leaves aromatic, narrowly elliptic to elliptic or occasionally obovate, with cuneate base, acute or occasionally obtuse apex and recurved margins, (3.5-)4-6(-9) cm long, (1.3-)1.5-2.5 cm wide, dark green and sparsely scaly above, greyish papillose beneath with small golden and larger brown scales. Petioles 8-10 mm long, scaly. Inflorescence terminal, rarely axillary in upper 1 or 2 nodes, (2-)4-6(-10)-flowered. Pedicels 1.3-2(-2.7) cm long, scaly. Calyx leafy with 5 conspicuously veined, ovate, acuminate lobes, 6-9(-11) mm long, scaly at base and on margins, with some loriform hairs and a tuft of hairs inside the apex of each lobe. Corolla pink, white flushed with pink or occasionally white, sometimes spotted within, campanulate to tubular-campanulate (1.8-)2-2.7(-3.2) cm long; tube (1-)1.3-1.7(-1.9) cm long, externally densely scaly, sometimes sparsely hairy also. Stamens 10, unequal; filaments 9-21 mm long, included or equalling corolla in length, densely pubescent, at least towards base. Ovary scaly with lobulate, glabrous basal disk. Style impressed, sharply deflexed and included (in var. *glaucophyllum*) or declinate and exserted (in var. *tubiforme*), glabrous. Stigma large, lobulate. Capsule ovoid, about 1 cm long, scaly, in persistent calyx.

Distribution. East Nepal, Sikkim, Bhutan.

Habitat. Forest clearings and on rocky slopes, 2700-3700 m.

Rhododendron glaucophyllum var. *glaucophyllum*
WALTER HOOD FITCH

RHODODENDRON CAMPYLOGYNUM

Subgenus *Rhododendron*
Section *Rhododendron* subsection *Campylogyna*

When Hutchinson wrote his account of this species for *Curtis's Botanical Magazine*, in 1935, he commented on the distinctness of series *Campylogynum* and then recorded the general doubt as to its relationships within the genus *Rhododendron*. Cullen, in his survey of subgenus *Rhododendron* (1980), has allied the species to *R. genesterianum* (subsection *Genesteriana*) and *R. pumilum* (subsection *Uniflora*). In Cullen's account it is the papillose leaves with distant scales and pruinose, often purplish, campanulate corollas of *R. genesterianum* which suggest its relationship to *R. campylogynum*, while the latter's habit of growth, inflorescence and fruit indicate a connection with *R. pumilum*.

Rhododendron campylogynum is extremely variable both morphologically and in hardiness, with numerous forms and varieties. Although some of these are horticulturally distinct, the differences are not sufficient to merit taxonomic status, even for the well-known var. *myrtilloides*, which has accordingly been reduced to synonymy. *Rhododendron campylogynum* is a deservedly popular plant in cultivation and between 1925 and 1975 the species received no fewer than seven awards at RHS shows. The size of this species and its liking for an open position make it a particularly good subject for the rock-garden, where it should be given a cool, moist root run. Here, too, its dense growth of neat, glossy leaves and nodding, thimble-shaped, pink to plum-purple, glaucous blooms, carried high above the foliage in May and June, show to advantage, especially when backlit by the sun. However, when selecting plants for the garden, it should be borne in mind that the altitudinal range (2400-4900 m) affects the hardiness of the various forms. The species can be propagated from cuttings, which will flower when between two and five years old.

Rhododendron campylogynum is common in the wild. It is found in the wettest areas of the Sino-Himalayan region, in India, Burma and China, and was first collected by Delavay in June 1884, above Tali on Mt Tsang-chan in west Yunnan, where it was growing at 3500 m and above. It was formally described by Franchet in the following year. George Forrest later collected the species several times on the eastern flank of the Tali range and it was Forrest who introduced the plant into cultivation in 1912, a year before Kingdon Ward brought seeds back from the Yunnan-Xizang border. Many of the forms in cultivation form neat cushions of growth only a few centimetres in height and one of these, the dwarf 'var. *myrtilloides*' is, despite its doubtful hardiness and tendency to die-back, one of the most desirable for the rock-garden. It was discovered by Kingdon Ward in 1914 in Upper Burma at about 4300 m and most of the plants originally grown in cultivation were raised from seeds collected by him in 1919. Kingdon Ward again gathered seed 20 years later, during the Vernay-Cutting expedition, after which Farrer and Cox also collected seed in the same area.

The plate painted by Lilian Snelling (tab. 9407) is of a plant grown at Lamellen, St Tudy, Cornwall by Mr Magor.

Rhododendron campylogynum Franchet in Bull. Soc. Bot. France 32: 10 (1885).
R. caeruleoglaucum Balf. fil. & Forr. in Notes Roy. Bot. Gard. Edinburgh 13: 34 (1920).
R. cremastum Balf. fil. & Forr., *op. cit.* 39.
R. glaucoaureum Balf. fil. & Forr., *op. cit.* 46.
R. charopoeum Balf. fil. & Forr., *op. cit.* 245 (1922).
R. damascenum Balf. fil. & Forr., *op. cit.* 254.
R. myrtilloides Balf. fil. & K. Ward, *op. cit.* 276 (1922).
R. cerasiflorum K. Ward in Gard. Chron. 93: 277 (1933), *nom. nud.*
R. rubriflorum K. Ward, Rhododendron Assoc. Year Book. 240 (1934), *nom. nud.*
R. campylogynum var. *celsum* Davidian in Rhododendron Year Book 8: 83 (1954).
R. campylogynum var. *charopoeum* (Balf. fil. & Forr.) Davidian, *loc. cit.*
R. campylogynum var. *cremastum* (Balf. fil. & Forr.) Davidian, *loc. cit.*
R. campylogynum var. *myrtilloides* (Balf. fil. & K. Ward) Davidian, *op. cit.* 84.

Description. An evergreen, prostrate or cushion-forming shrublet, up to 60 cm in height, or less commonly an erect shrub attaining 1-2 m in height; young growth scaly and often pubescent, the older branches rough and peeling. Leaves coriaceous, dark green above, often silvery beneath, congested, obovate, oblanceolate or elliptic, (10-)14-25(-34) mm long, (4-)7-13 mm wide, with cuneate, decurrent base and obtuse or occasionally subacute apex, margins recurved, sparsely pubescent along midrib above, papillose and very sparsely scaly beneath, the scales caducous. Petioles about 2 mm long. Inflorescence terminal, 1-2(-4)-flowered, the flowers nodding. Pedicels usually tinged with red, rigid, accrescent, 2.5-5 cm long (7 cm long in fruit). Calyx leafy, deeply lobed; lobes oblong-obovate, 4-7 mm long, usually without scales or hairs. Corolla fleshy, pink, red, crimson or purple, bell-shaped, (10-)13-20(-23) mm long, externally pruinose, glabrous; tube 7-13 mm long, pubescent within near base. Stamens 8-10, unequal; filaments pubescent in lower third to half. Ovary laxly scaly with lobed, glabrous, basal disk. Style stout, curved downwards, glabrous. Stigma clavate. Capsule ovoid, 7-9 mm long, sparsely scaly with persistent calyx.

Distribution. North India (Arunachal Pradesh), north-east Burma, China (Yunnan, south and south-east Xizang) .

Habitat. Cliffs, moorland and in scrub, 2400-4250(-4900) m.

Rhododendron campylogynum
LILIAN SNELLING

RHODODENDRON GENESTERIANUM

Subgenus *Rhododendron*
Section *Rhododendron* subsection *Genesteriana*

In a remote corner of Yunnan, at Kieunatong, Père A. Genestier regularly advised and assisted explorers of the area, and George Forrest named this rhododendron after him in recognition of his help.

In comparison with other rhododendrons, *R. genesterianum* can be termed 'curious'. It is certainly different, with purple young growth, willow-like leaves and loose racemes of long-stalked, nodding, bell-shaped flowers very dark plum-red/purple in colour with a dense external bloom. The species was introduced by Forrest after he discovered it on the Salween-Irrawaddy divide in April 1919 (F 17824) and re-introduced several times subsequently by him and by others. He liked the plant, as did Kingdon Ward, who nevertheless saw it as interesting rather than beautiful. Farrer on the other hand considered it to be 'almost ugly'. Opinions continue to vary: its popularity is not ever likely to be universal and it remains rare in cultivation, but the species is an acquired taste, and a flourishing plant in full bloom soon wins its owner's approbation. The species is, rather unfortunately, too tender for most British gardens; even where it succeeds, it is not an easy subject, but is distinctly worth trying. The flowers are produced in April and May in some profusion.

Tab. 9310 in *Curtis's Botanical Magazine* was painted by Lilian Snelling in 1928 from a plant raised by the Marquis of Headfort at Kells, Co. Meath, Eire, grown from seed of Forrest's no. 19117, which was collected in 1921 at Tsarong in Xizang. Originally included in series *Glaucum*, *R. genesterianum* has been transferred by Cullen to subsection *Genesteriana* which includes parts of series *Glaucum* and series *Glaucophyllum* subseries *Genesterianum*. This subsection appears to be most closely related to subsection *Campylogyna*, despite the obvious discrepancies of habit, i.e. comparative plant height and leaf size.

Rhododendron genesterianum Forr. in Notes Roy. Bot. Gard. Edinburgh 12: 122 (1920).
R mirabile K. Ward in Gard. Chron. 92: 465 (1932) *nom. nud.*

Description. A lax, evergreen, terrestrial shrub, up to 5 m in height with smooth, purplish bark and glabrous young growth. Leaves thin in texture, dark green, crowded at the ends of the branches, narrowly elliptic to narrowly oblanceolate or lanceolate, 6.5-12(-15) cm long, (1.2-)2.5-4.5 cm wide, with tapered base and abruptly acuminate apex; lower surface conspicuously white-papillose with distant, small, golden to brown scales. Petioles 0.5-2 cm long, sparsely scaly. Inflorescence a loose, terminal (4-)10-12(-15)-flowered raceme; rachis 1-1.5 cm or more, long. Pedicels slender, more or less erect, (1.6-)2-3 cm long, pruinose and scaly. Calyx obsolete, an undulate or shallowly lobed, glabrous rim 1-2 mm long, pruinose, sometimes sparsely scaly. Corolla fleshy, dark reddish purple and pruinose, tubular-campanulate, 1.2-1.7 cm long; tube 6-8 mm long, glabrous, with more or less erect lobes. Stamens (8-)10, unequal; filaments glabrous. Ovary ovoid or conoid, truncate, glaucous, scaly with green or purple, lobulate basal disk. Style stout, 7-9 mm long, sharply bent, thickened to apex. Stigma discoid, lobulate. Capsule cylindrical, 6-9 mm long, scaly with persistent calyx.

Distribution. China (West Yunnan, south-east Xizang), north Burma.

Habitat. Scrub, cane brakes, forest margins, or in open thickets, 2450-4250 m.

Rhododendron genesterianum
LILIAN SNELLING

RHODODENDRON LOWNDESII

Subgenus *Rhododendron*
Section *Rhododendron* subsection *Lepidota*

This attractive slow-growing dwarf species is not an easy plant to grow, needing good drainage and preferring a north-facing site with plentiful light. It dislikes damp and frost which soon cause it to die back and for this reason it is best grown in the rock-garden or a peat wall, which situations most closely emulate the rock ledges and crevices of its natural habitat. Nurseryman Peter Cox grows the species with varying success at Glendoick and has found that protecting it with a cloche from September to May prevents frosted or rotted shoots. Despite the extra trouble needed, *R. lowndesii* is a species well worth growing and in favourable conditions will flower freely on mats of compact, low, stoloniferous growth. The flat-faced, pale yellow, crimson-spotted flowers are individually short-lived, but are large for the size of the plant and are borne well above the foliage on erect, solitary or paired pedicels from May to June. *Rhododendron lowndesii* is closely related to the type species of subsection *Lepidota*, *R. lepidotum*, differing mainly in its creeping habit and deciduous bristly leaves (*R. lepidotum* is evergreen); overall, it is more hairy than *R. lepidotum*. The subsection as a whole is allied to subsections *Baileya*, *Trichoclada* and *Lapponica*. The plant was named for Colonel D.G. Lowndes (1899-1956) who discovered it in 1950 in the Marsiandi Valley of Nepal, where it was growing at an altitude of 4050 m. George Smith of Manchester University and Stainton, Sykes & Williams subsequently collected it during the 1950s and the species was named and described by David Davidian in 1952, the same year that Polunin, Sykes & Williams collected the species in Nepal (no. 3846). A plant raised from seed taken from their herbarium specimen, flowered for the first time at the Royal Botanic Garden, Edinburgh, in 1956. The accompanying plate was painted by Mary Mendum (Mary Bates), especially for this volume, at Edinburgh in 1990.

Rhododendron lowndesii Davidian in Notes Roy. Bot. Gard. Edinburgh 21: 99 (1952).

Description. A dwarf, prostrate and creeping shrub up to 25(-30) cm in height, branches slender, pubescent and with some loriform hairs, somewhat scaly. Leaves deciduous, chartaceous, bright green above, paler beneath, elliptic to oblanceolate or obovate, (9-)15-25(-28) mm long, (4-)6-12 mm wide, with tapering, decurrent base and rounded, mucronate apex, margins loriform-ciliate, upper surface pubescent and sparsely scaly, lower surface with distant, translucent, yellow scales. Petioles 2-3 mm long, ciliate. Inflorescence terminal, 1-2-flowered. Pedicels slender, somewhat accrescent in fruit, 15-20(-43) mm long, usually sparsely bristly and scaly. Calyx green or stained red, 5-lobed, lobes oblong-ovate, rounded, 2.5-3.5(-5) mm long, externally sparsely scaly, margins ciliate. Corolla yellow, sometimes freckled with red, campanulate to rotate, 13-15(-17) mm long overall; tube very short, externally sparsely to densely scaly; lobes 5, spreading widely. Stamens 10, exserted from tube, 6-10 mm long; filaments densely pubescent for up to two-thirds of their length. Ovary conical, densely scaly. Style short, stout, strongly curved, glabrous. Capsule cylindrical or conical, 4-5(-7) mm long, more or less scaly and bearing the persistent calyx.

Distribution. Nepal.

Habitat. Rock crevices, ledges and peaty banks, 3800-4550 m.

Rhododendron lowndesii
MARY MENDUM

RHODODENDRON BAILEYI

Subgenus *Rhododendron*
Section *Rhododendron* subsection *Baileya*

Rhododendron baileyi is one of the small, scaly, 'flat-flowered' rhododendrons previously classified as a member of the series *Lepidotum*. Recently the species has been re-classified by Cullen as the type species of the currently monotypic subsection *Baileya*. He points out that despite its obvious close relationship with subsection *Lepidota* (and especially with *R. lepidotum*) *Rhododendron baileyi* must be considered distinct by reason of its crenulate scales and sharply deflexed style. Similar crenulate scales, rather like minute brown daisies, are found in subsection *Saluenensia*, but in most other characters, that subsection differs considerably from subsection *Baileya*. In 1913, *R. baileyi* was discovered by Capt. (later Lt.-Col.) F.M. Bailey in the upper valley of Nyanjang, Xizang, near the north-eastern border with Bhutan. Seeds were collected and sent with those of other rhododendrons to the Royal Botanic Garden at Edinburgh for re-distribution to other gardens.

Though a rather untidy plant best grown behind true dwarf species, *R. baileyi* is an attractive woodland shrub, with the best forms freely producing widely open, brilliant, sometimes speckled, rosy purple blooms which are relatively large for a plant attaining only about 1 m in height. As many as 18 flowers have been counted in one inflorescence, but five to nine are more usual. The species is hardy, although young growth and flower-buds are occasionally killed by frost, and in his account of the species for *Curtis's Botanical Magazine*, Stapf claims that it is also drought-resistant, at least, it has proved to be so in Cornwall. When not in flower, the coloration of the encrusted scales on twigs and lower leaf-surfaces provides secondary interest. Propagation is best effected from seed, as cuttings are difficult to root. A plant exhibited by A.C. and J.F.A. Gibson of Glenarn, Dumbartonshire in 1960, received an Award of Merit, the only award this lovely plant has, as yet, received from the RHS.

Dame Alice Godman's garden at Horsham supplied the material from which most of Lilian Snelling's plate was drawn; figs. 5, 9 and 10 were drawn from specimens supplied by J.C. Williams of Caerhays Castle, Cornwall, whose plant was used by Sir Isaac Bayley Balfour to write the original description of the species. Further material was received from H. Armytage-Moore of Roswallen, Co. Down.

Rhododendron baileyi Balf. fil. in Notes Roy. Bot. Gard. Edinburgh 11: 23 (1919).
R. thyodocum Balf. fil. & Cooper in Notes Roy. Bot. Gard. Edinburgh 11: 148 (1919).

Description. A small shrub up to 1 m in height with peeling bark and pseudowhorls of usually 3 or 4, slender, stiffly erect, densely scaly branches at the end of each year's growth. Leaves more or less coriaceous, in clusters of 3-7 at upper nodes, narrowly elliptic to elliptic, occasionally obovate or ovate, (21-)30-50 mm long, (8-)14-19(-33) mm wide, with usually cuneate base and obtuse to rounded apex, upper surface dark, dull green, densely scaly, but scales caducous, lower surface brownish cinnamon or dark brown with dense, scurfy, overlapping, persistent crenulate scales. Petioles 5-10 mm long, scaly. Inflorescence terminal (4-)5-8 (rarely 18)-flowered; rachis elongate, up to 17 mm long, pubescent. Pedicels red, slender, 12-22(-35) mm long, scaly. Calyx green or red with 5, more or less deltoid, often unequal lobes, (1.5-)2-4 mm long, scaly, often ciliate with loriform setae. Corolla red, magenta or purple, often speckled with darker flecks or spots, rotate, about 3 cm across, with very short, broad tube, 5-7(-9) mm long, glabrous or finely pubescent within, tube and lobes externally scaly. Stamens 10, subequal; filaments rosy purple, 8-10 mm long, variously white-pubescent, some near the middle only, others almost to apex. Ovary cylindrical-conoid with rounded apex, ribbed, densely scaly and with green, 5-lobulate, glabrous basal disk. Style red, stout, 3-6 mm long, curved downwards, dilated immediately beneath stigma, glabrous. Stigma capitate, lobulate. Capsule 5-7.5 mm long, scaly.

Distribution. North-east India (Sikkim), Bhutan and China (south Xizang).

Habitat. Forest, hillsides, sometimes on scree or among rocks, 3050-4250 m.

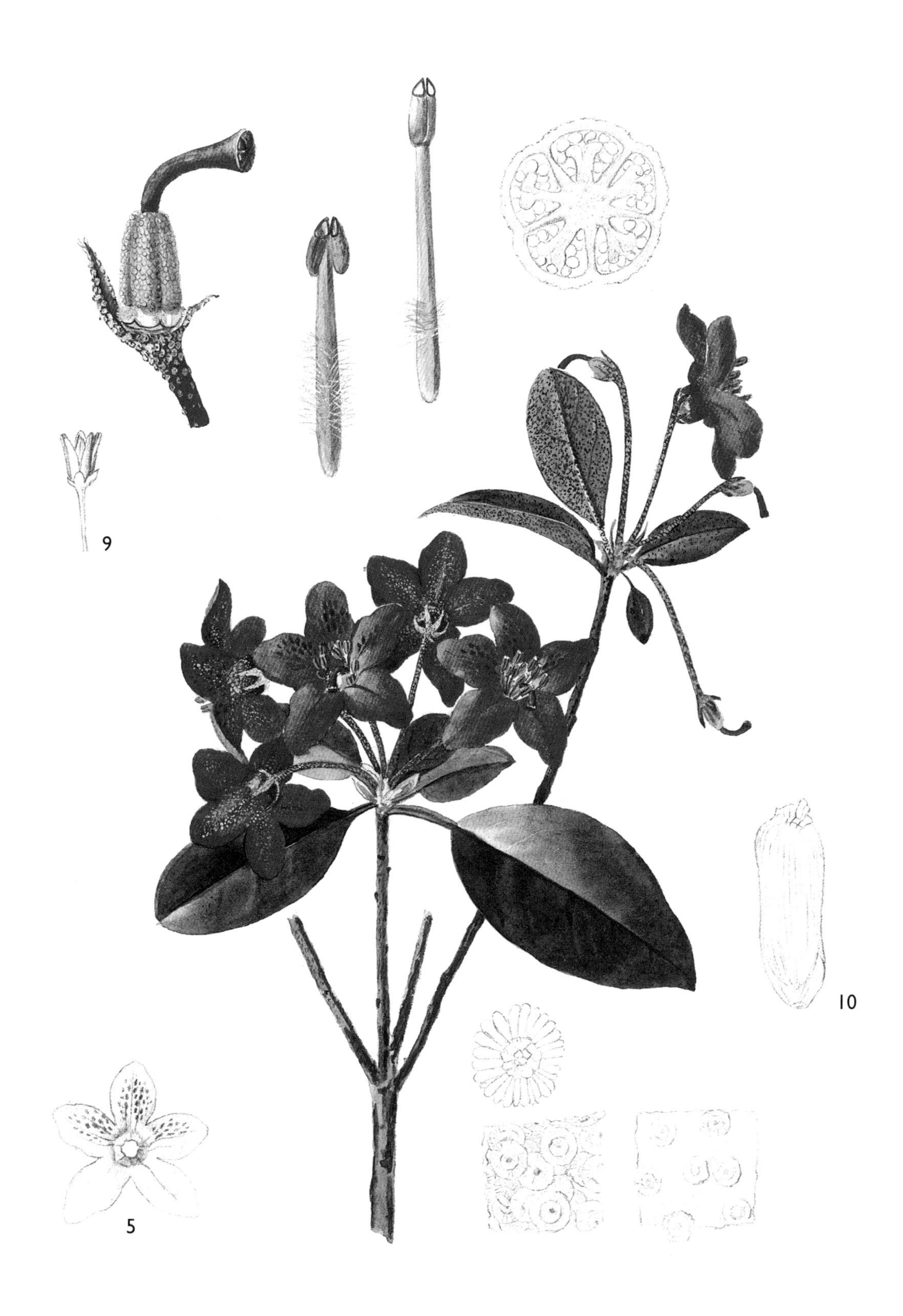

Rhododendron baileyi
LILIAN SNELLING

RHODODENDRON TRICHOCLADUM

Subgenus *Rhododendron*
Section *Rhododendron* subsection *Trichoclada*

Many growers would consider this hardy and precociously flowering species unworthy of a place in the garden. Undoubtedly the main feature is the flower colour, a pallid lemon-yellow, usually with the upper lobes spotted green, and this is certainly not a combination to command attention. Nor is the growth habit — when in bloom the plants exhibit an unattractive mode of growth, producing stiff, slender, erect branches topped by a tuft of tiny young leaves, initially also held stiffly erect. Later, when the leaves are fully expanded, this awkward habit is concealed. But even the most unattractive plant will find supporters, and *R. trichocladum* is not completely unworthy. In groups in a sunny spot, frequently with bronzed young growth, the modest display can be attractive, often lasting from April to July, with a second crop of flowers in the autumn, when the leaves turn yellow.

Rhododendron trichocladum is included here as a representative of subsection *Trichoclada* (previously classified as series *Trichocladum*) which in Cullen's revision includes *R. lepidostylum*, *R. mekongense* and *R. caesium*, all very closely related to each other and all of which produce terminal inflorescences. Plants of *R. trichocladum* which produce both terminal and axillary inflorescences, frequently prove to be hybrids with *R. racemosum*. Such crosses are known to occur in the wild. Abbé Delavay first discovered *R. trichocladum* in June 1885 in China, on the slopes of Cangshan in the Tali range where it is very common. It was growing at an elevation of 2500 m. George Forrest, working for A.K. Bulley, collected it in the same area at the somewhat higher altitude of 2700-3000 m in 1906. However, the first successful introduction was said to be made several years later in 1910, again by Forrest who gathered seed many times subsequently, and plants raised from his introductions can still be seen at Windsor, Nymans and Borde Hill. Rock and Kingdon Ward too, introduced *R. trichocladum* several times. The collection KW 8259, about which there was some doubt, has now been confirmed by Cullen as belonging to this species.

The Windsor plant received an Award of Merit from the RHS in 1971 when it was exhibited as *R. lophogynum* by the Crown Estate Commissioners. Mr G.W.E. Loder, Wakehurst Place, supplied the material from which tab. 9073 was painted by Lilian Snelling in 1924.

Rhododendron trichocladum Franchet in Bull. Soc. Bot. France 33: 234 (1886).
R. xanthinum Balf. fil. & W.W. Sm. in Trans. Bot. Soc. Edinburgh 27: 87 (1916).
R. lithophilum Balf. fil. & K. Ward in Notes Roy. Bot. Gard. Edinburgh 13: 275 (1922).
R. oulotrichum Balf. fil. & Forr. in Notes Roy. Bot. Gard. Edinburgh 13: 281 (1922).
R. lophogynum Balf. fil. & Forr. ex Hutch. in Stevenson, Sp. Rhodod. (1930) *nom. nud.*

Description. A deciduous or semi-evergreen shrub up to 1.5(-2.4) m in height, with slender, scaly and bristly branches held stiffly erect; young growth scaly and usually more or less densely loriform-setose, the setae twisted or curled. Leaves deciduous or somewhat more persistent, coriaceous, oblong, obovate or obovate-elliptic, (16-)24-40(-56) mm long, (6-)10-12(-26) mm wide, with cuneate base and more or less rounded, mucronate apex, margins loriform-ciliate, upper surface dull green with straight loriform setae and/or puberulent, sometimes scaly; lower surface more or less twisted loriform-setose. Petioles 2-5 mm long, scaly and loriform-setose. Inflorescence a terminal, umbellate cluster of 1-3(-5) precocious flowers. Pedicels 8-13(-35) mm long, scaly, loriform-setose. Calyx with 5, often unequal, lanceolate, scaly and loriform-ciliate lobes 2-5 mm long. Corolla pale yellow, greenish or reddish yellow, widely funnel-shaped, (15-)18-23 mm long, up to 45 mm across lobes; tube 8-11 mm long, externally scaly and variably loriform-setose; lobes oblong with rounded apex. Stamens 10, unequal, 9-12(-17) mm long, exserted from the throat, but shorter than the corolla-lobes; filaments pubescent towards base. Ovary conoid, scaly, occasionally with a few loriform setae at the apex. Style stout, glabrous, or puberulent at base, usually sharply bent, almost right-angled. Capsule ellipsoid or cylindrical, (6-)8-10 mm long, scaly and sometimes with a few persistent setae.

Distribution. China (central and south-west Yunnan), north Burma.

Habitat. Slopes, cliffs and other rocky places, and in scrub, boggy places and along river banks, 2450-3350 m.

Rhododendron trichocladum
LILIAN SNELLING

RHODODENDRON MEKONGENSE var. MELINANTHUM

Subgenus *Rhododendron*
Section *Rhododendron* subsection *Trichoclada*

Until Cullen reduced it to synonymy under *R. mekongense* in 1978, this plant was accepted at specific rank as *R. melinanthum*. Then, in his revision of subgenus *Rhododendron* (1980) Cullen confirmed his re-classification and accepted four varieties within *R. mekongense*, (vars. *mekongense*, *melinanthum*, *rubrolineatum* and *longipilosum*). The species remains in series *Trichocladum* Balf. fil., which in the new revision is re-classified as subsection *Trichoclada*. *Rhododendron mekongense* is very like the type species of the subsection, (*R. trichocladum*) but differs in indumentum, scales (equal size in *R. trichocladum*, unequal in *R. mekongense*) and pedicel length as well as in geographical distribution (*R. trichocladum* occurs in north Burma, central and south-west Yunnan; *R. mekongense* is found in Nepal, north-east Burma, north-west Yunnan and south-east Xizang). Var. *melinanthum* is distinguished by the lack of loriform hairs on the calyx and pedicels, although occasionally they are present at the base of the pedicels.

The plant originally described as *R. melinanthum* was discovered in June 1913 by Kingdon Ward on the upper part of the Mekong-Salween divide (KW 406) in a valley where, at altitudes between 3600 and 4000 m, it grew as undergrowth in *Abies* forest. He collected seed later in the year from which the plant depicted in tab. 8903 by Lilian Snelling, was raised at the Royal Botanic Garden, Edinburgh. Growing in the garden at Borde Hill, Sussex, is a plant introduced as *R. melinanthum* KW 406, which was later identified as *R. mekongense* var. *mekongense*. It is an attractive form of the species with bronze young growth and rich yellow flowers in April. Var. *melinanthum* is the largest and showiest member of the subsection and flowers more freely if planted in an open situation; the leaves are deciduous, and the flowers precocious, and usually bright yellow (although the colour varies), and appearing in April and May, but no member of the subsection can be said to be beautiful. This does not mean that the group should be excluded from collections, merely that these species should be thoughtfully planted among or beside other equally subdued species where their quiet appeal and softer colouring can be properly appreciated.

As *R. mekongense*, a plant of the type number of this species, KW 406, received an Award of Merit in 1979 when exhibited at the RHS by R.N. Stephenson Clarke.

Rhododendron mekongense Franchet in J. de Bot. 12: 263 (1898). var. **melinanthum** (Balf. fil. & K. Ward) Cullen in Notes Roy. Bot. Gard. Edinburgh 36: 115 (1978).
R. melinanthum Balf. fil. & K. Ward in Trans. Bot. Soc. Edinburgh 27: 85 (1916).
R. chloranthum Balf. fil. & Forr. in Notes Roy. Bot. Gard. Edinburgh 12: 98 (1920).
R. semilunatum Balf. fil. & Forr., *op. cit.* 13: 292 (1922).

Description. An erect, much-branched shrub up to 2 m in height, flowering precociously and with scaly and loriform-setose young growth. Leaves more or less deciduous, coriaceous, obovate or obovate-elliptic, 2.5-4.5(-6.5) cm long, 1.4-2.1(-2.7) cm wide with cuneate base and more or less rounded apex. Petioles 2-4 mm long, loriform-setose. Upper surface pubescent along midrib and occasionally scaly; lower surface somewhat glaucous, variably loriform-setose, densely scaly, the scales very unequal in size. Inflorescence a terminal 2-4(-5)-flowered cluster. Pedicels (1.1-)1.5-2.2 cm long, variably scaly and not loriform-setose, or if so, then only at base. Calyx obscurely lobed; lobes up to 2.5 mm long, sometimes one much longer, up to 7 mm long, scaly. Corolla yellow or greenish, sometimes stained red, with or without spots, widely campanulate or campanulate, 5-lobed, 1.7-2.3 cm long, externally scaly. Stamens 10, unequal, exserted from throat of tube, 0.8-1.7 cm long; filaments variably pubescent. Ovary densely scaly. Style equalling or exceeding corolla in length, glabrous or sometimes puberulent at base. Capsule cylindrical, 9-11 mm long, densely scaly.

Distribution. Var. *melinanthum* is apparently restricted to a small area of north-east Upper Burma. The species as a whole grows over a much wider area, including Nepal, north-east Burma, north-west Yunnan and south-east Xizang.

Habitat. Margins of conifer and rhododendron forests and other thickets, 3600-4200 m. The altitudinal range of the species as a whole is somewhat extended from 2900 to 4400 m.

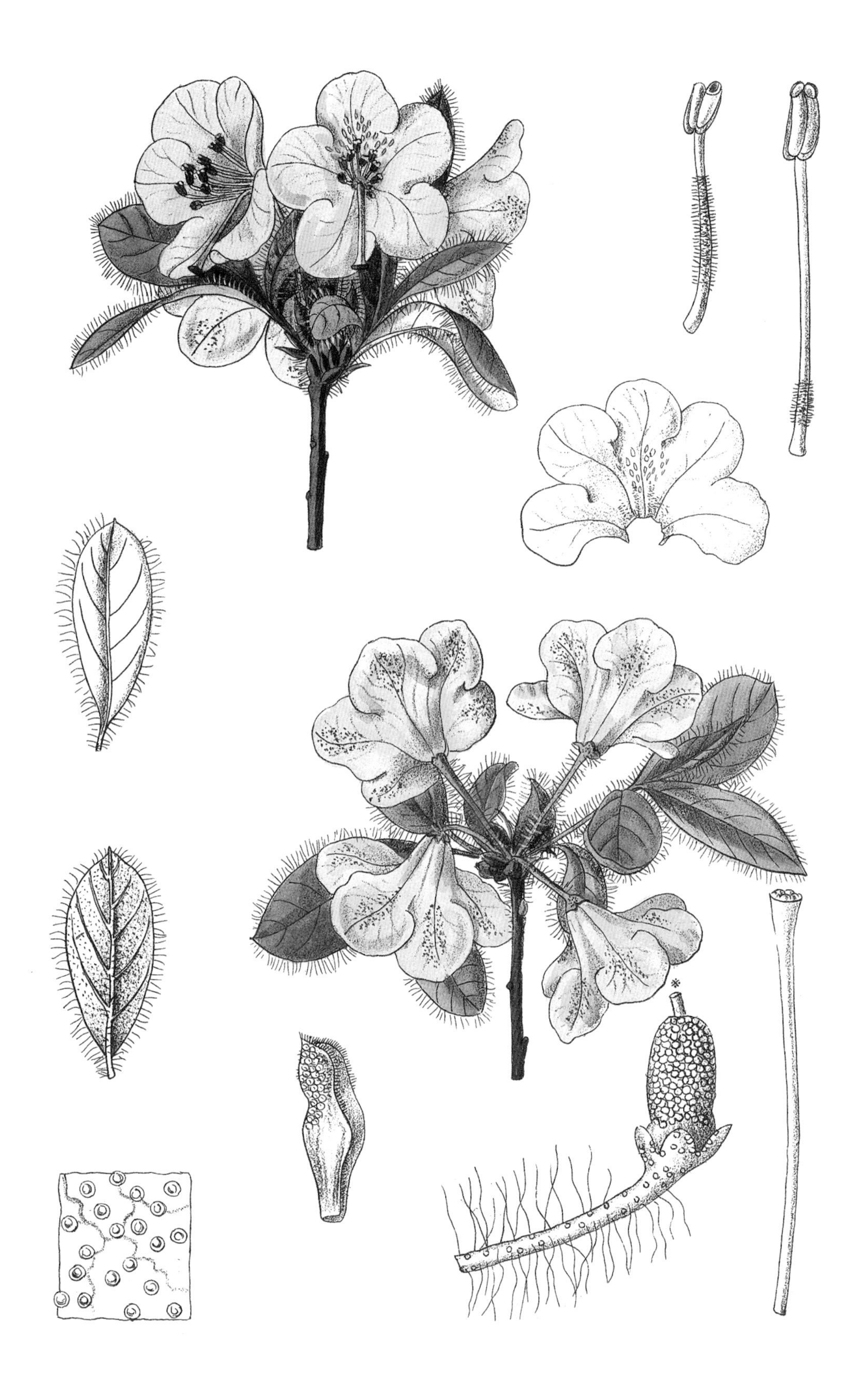

Rhododendron mekongense var. *melinanthum*
LILIAN SNELLING

RHODODENDRON AFGHANICUM

Subgenus *Rhododendron*
Section *Rhododendron* subsection *Afghanica*

This species is so distinctive that it was given a subsection of its own in Cullen's recent (1980) revision. Previously, with *R. hanceanum*, it comprised the subseries *Hanceanum* of series *Triflorum*, but as Cullen has noted, not only did the two species have little in common, but the affinities of *R. afghanicum* are more likely to be with subsections *Boothia* and *Camelliiflora*, than with subsection *Triflora*. Davidian holds a different view, preferring to retain this species in subseries *Hanceanum*, while acknowledging that it does not fit well into that or any other current group. However *R. afghanicum* is classified, its lack of close relatives makes it a botanically interesting, if not an ornamental plant. It is a slow-growing dwarf species, evergreen and semi-prostrate, but not completely hardy and therefore suited only for sheltered sites in mild areas, where it will flower in June and July and readily set seed. *Rhododendron afghanicum* was discovered by Aitchison when, as official botanist, he accompanied Lord Roberts on his expedition to the Kurrum Valley in 1879, and it was the first *Rhododendron* species to be found in Afghanistan. Aitchison was responsible for its introduction into cultivation in 1880 and a few plants raised from the seed he collected were still extant in 1946. Soon after the Second World War, the knowledge that the species is very poisonous to livestock and fatal to sheep, caused the surviving plants from Aitchison's introduction, then growing at Tower Court, Ascot, to be destroyed. The species was not re-introduced until 1969, when Hedge and Wendelbo collected seed.

Tab. 8907 was painted by Lilian Snelling from a plant which flowered in the Royal Botanic Garden, Edinburgh, some years before the plate was published in *Curtis's Botanical Magazine*. The reason for the delay was the gap of eighteen years between the publication of volume 146 (1921) and volume 147. The latter, in which tab. 8907 is included, was not published until 1938.

Rhododendron afghanicum Aitch. & Hemsley in J. Linn. Soc., Bot. 18: 75 (1880).

Description. A low, evergreen, often straggly shrub up to 50 cm in height; young growth scaly and sometimes minutely hairy. Leaves coriaceous, thick, narrowly elliptic to elliptic, (30-) 47-80 mm long, 13-25 mm wide, with rounded to cuneate base and a more or less obtuse apex, upper surface dark green without scales, but with the midrib minutely hairy at the base and on the petiole, lower surface pale green, scaly. Petioles about 10 mm long, sparsely scaly. Inflorescence a more or less loose terminal raceme of (5-)12-16 flowers; rachis 2-5 cm long. Pedicels 5-15 mm long, densely scaly. Calyx-lobes unequal, 4-6 mm long, narrowly triangular to oblong, apex acute or rounded, externally scaly and with scaly margins. Corolla white or greenish, campanulate; tube 6-8 mm long, sparsely hairy within; lobes spreading, about 5 mm long, externally glabrous. Stamens 10, unequal, exserted; filaments hairy towards base. Ovary densely scaly with minutely hairy basal disk. Style impressed, sharply deflexed, glabrous. Stigma large, lobulate. Capsule cylindrical, about 7 mm long, scaly.

Distribution. Known only from the Kurrum Valley, which crosses the Afghanistan-Pakistan border.

Habitat. Cliffs and forests, creeping among rocks (usually gneiss or limestone), 2000-3000 m.

Rhododendron afghanicum
LILIAN SNELLING

RHODODENDRON TRICHOSTOMUM

Subgenus *Rhododendron*
Section *Pogonanthum*

The Abbé Delavay first found this plant in May 1887 near Lankiung, which is to the north of Tali in western Yunnan, China, where it grows on the margins of forests and thickets, on open slopes and in scrub at altitudes ranging from 2500 to 4600 m. Franchet described the species from Delavay's material in 1895, but it is not thought to have been introduced into cultivation until 1908 when E.H. Wilson collected it in western Sichuan (no. 1328). Forrest, Kingdon Ward and Rock all re-introduced the species in later years, and it was a plant raised from Forrest's seed which first flowered in cultivation in 1917.

Although more lax in growth than most of the other members of the section, from which it is readily distinguished by its narrow leaves, *R. trichostomum* is also one of the most ornamental of them and has received no less than six Awards at RHS shows since its introduction, if not always under its correct name. They were as follows: A.M. 1925 when exhibited by A.K. Bulley, Neston (as *R. ledoides*); A.M. 1960 to the clone 'Sweet Bay', exhibitors Crown Estate Commissioners, Windsor (from F 20480); A.M. 1971 to the clone 'Quarry Wood' and A.M. 1972 (as *R. trichostomum*), exhibitors Mr and Mrs Martyn Simmonds, Quarry Wood, Newberry; A.M. 1972 to the clone 'Lakeside', exhibitors Crown Estate Commissioners, Windsor; F.C.C. 1976 to the clone 'Collingwood Ingram', exhibitor Lady Anne Palmer, Rosemoor Garden, Torrington, Devon.

The species flowers late in the season, in May and June, and being variably hardy should be given a sheltered though sunny position. Wilson reported a plant from Tatsienlu as being about 2 m in height, but in cultivation plants rarely exceed 1 m. The species flowers freely, the rounded white or pink trusses and small leaves resembling a miniature replica of its larger hardy relatives. Plants are best raised from seed as cuttings do not root readily.

Cullen (1980) noted that section *Pogonanthum* appears to be allied to section *Rhododendron* subsections *Lapponica* and *Rhododendron*. *Rhododendron ledoides* and *R. trichostomum* var. *ledoides* are now included in the synonomy of *R. trichostomum*, and *R. hedyosmum* is variously treated as a synonym of *R. trichostomum* or, by some authorities, as a variety of it. Cullen (1980) suggested that *R. hedyosmum* is probably a hybrid. Balfour originally gave the name to a large-flowered plant raised at Edinburgh from seed of Wilson's no. 1208, which flowered and set seed in 1916 (the type collection of *R. sargentianum*) and that plant is known only in cultivation, if indeed it remains in cultivation, which is doubtful. The plant depicted in tab. 9202 of *Curtis's Botanical Magazine* was raised from Wilson's seed, but it is not *R. hedyosmum*, nor is it *R. trichostomum*: the calyx is much too large for either plant. The species was also figured in tab. 8831 as *R. ledoides* and in tab. 8871 as *R. sargentianum*. The present plate (plate 57) was prepared by Valerie Price for this book from material grown at Wakehurst Place, Sussex.

Rhododendron trichostomum Franchet in J. de Bot. 9: 396 (1895).
R. fragrans sensu Franchet, Bull. Soc. Bot. France 34: 284 (1887) *non* (Adams) Maxim. (1870).
R. ledoides Balf. fil. & W.W. Sm., Notes Roy. Bot. Gard. Edinburgh 9: 243 (1916).
R. radinum Balf. fil. & W.W. Sm., *op.cit.* : 268.
R. sphaeranthum Balf. fil. & W.W. Sm., *op. cit.* : 278.
R. trichostomum var. *ledoides* (Balf. fil. & W.W. Sm.) Cowan & Davidian in Rhododendron Year Book 2: 84 (1947).
R. trichostomum var. *radinum* (Balf. fil. & W.W. Sm.) Cowan & Davidian, *loc. cit.*

Description. A sometimes lax, but often compact and rounded, much-branched, evergreen, aromatic shrub up to 1(-1.5) m in height. Leaves coriaceous, linear, linear-lanceolate, oblong or oblanceolate, 12-30(-34) mm long, 3-5(-8) mm wide, at least 4 times longer than wide, tapered to the base, more or less rounded at the mucronate or emarginate apex, margin revolute, upper surface dull green, scaly or glabrous, lower surface brown, due to dense, overlapping flaky scales in 2 or 3 layers, those of the lowest layer golden, paler than the rest. Petioles about 3 mm long, densely scaly. Inflorescence terminal, more or less globose, with 8-20 densely packed flowers. Floral bud-scales often persistent. Pedicels up to 7 mm long, scaly, sometimes also puberulent, occasionally completely devoid of indumentum. Calyx 1-2.5 mm long, 5-lobed, lobes oblong or narrowly triangular, usually externally scaly with loriform-ciliate margins, variably minutely pubescent within. Corolla white or pink, fading with age, hypocrateriform, 6-11.5(-20) mm long overall; tube often deep pink, 4.5-8(-10) mm long, externally glabrous, pilose in the throat; lobes usually externally laxly scaly. Stamens 5(-6), unequal, 3-6 mm long, shorter than the tube. Ovary scaly. Style stout, short, glabrous. Stigma turbinate. Capsule 2-4 mm long, scaly.

Distribution. China (north & north-west Yunnan, south-west & central Sichuan).

Habitat. Edges of forest and thickets, scrub, open slopes, 2500-4600 m.

Rhododendron trichostomum
VALERIE PRICE

RHODODENDRON CALOPHYTUM var. CALOPHYTUM

Subgenus *Hymenanthes*
Section *Ponticum* subsection *Fortunea*

The large leaves and ventricose corollas of this species suggest affinities with subsection *Grandia*, but *R. calophytum* is in fact a member of subsection *Fortunea*, to which it is more closely allied. The species comprises two subspecies, distinguished by leaf-size and the number of flowers in an inflorescence. In both characters, var. *calophytum* exceeds var. *openshawianum*. The latter variety has been reduced by Chamberlain from specific rank; it is not known to be in cultivation. Although too large for many gardens, averaging 12 m in height, *R. calophytum* is a magnificent species for gardens where space is no problem. It is easily grown and long-lived, merely requiring some shelter from wind to flourish. Once the flowering stage has been reached (after some years) flowers are freely produced early in the season from February or March to April, but the species is completely hardy and frost damage is usually minimal. A notable feature is the very large stigma, by which the species is instantly, if somewhat unscientifically, recognizable.

The species was discovered by Père (Abbé) G. David in 1869. He found it growing near the French Mission Station, Moupine in west Sichuan. Subsequently, E.H. Wilson found the species in other parts of that country, including Tanlanshan to the north of Moupine, the forests south east of Tatsienlu, and Washan. The species was introduced into cultivation in 1904 when Wilson was collecting for Veitch and he re-introduced it four years later, when on an expedition from the Arnold Arboretum. He reported it to be common in the forests of west Sichuan and the largest rhododendron in that region. The species flowered for the first time in cultivation in 1916 for Messrs Reuthe, and when exhibited by them in March 1920, received an Award of Merit from the RHS. The award was for a form with pink-flushed white flowers. Dame Alice Godman of South Lodge, Lower Beading, Sussex, grew a lovely pale pink form which, when exhibited in April 1933, was awarded a First Class Certificate. The material from which tab. 9173 was painted by Lilian Snelling, was supplied by Mr E. J.P. Magor, Lamellen, St Tudy, Cornwall.

Rhododendron calophytum Franchet in Bull. Soc. Bot. France 33: 230 (1886). var. **calophytum**.

Description. An evergreen tree up to 12 m in height with rough bark and stout branches. Leaves coriaceous, oblong-oblanceolate, 14-30 cm long, 4-7.2(-8)cm wide, with cuneate base and cuspidate to acuminate apex, both surfaces glabrous except for persistent vestiges of juvenile floccose tomentum along the midrib beneath. Petioles stout, 1-2 cm long, glabrous or thinly floccose. Inflorescence a lax truss of 15-30 flowers; rachis 1.2-2.5 cm long. Pedicels red, 3-5.5 (-7) cm long, glabrous. Calyx a minute, undulate rim, about 1 mm long, glabrous. Corolla white to pink, white flushed with pink (or pale lilac-pink flushed with purple in cultivation) with crimson to purple flecks and basal blotch, open-campanulate, ventricose below, 5-6 cm long, 5-7-lobed. Stamens 15-20, unequal, included; filaments puberulent near base. Ovary glabrous. Style stout, glabrous. Stigma yellow-green, sometimes flushed with red, discoid, 8 mm in diameter. Capsule 2.5-3.3 cm long.

Distribution. Both varieties are found in China (central & east Sichuan, north-east Yunnan)

Habitat. Forests and thickets, 1800-4000m.

Rhododendron calophytum var. *calophytum*
LILIAN SNELLING

RHODODENDRON DECORUM subsp. DIAPREPES

Subgenus *Hymenanthes*
Section *Ponticum* subsection *Fortunea*

Hutchinson's text in *Curtis's Botanical Magazine* which accompanies tab. 9524 begins: '*Rhododendron diaprepes* is so very near the better known *R. decorum* Franch. that one might be quite justified in regarding it as a variety or geographical race of that species'. Chamberlain, in his 1982 revision of subgenus *Hymenanthes*, having retained *R. diaprepes* at specific rank and subsequently examined further material, agreed with Hutchinson's opinion. In a footnote to his description of the species, Chamberlain reduced *R. diaprepes* to the rank of subspecies and made the necessary new combination. Certainly, apart from somewhat differing geographical ranges, the main distinction between the two subspecies is one of leaf and flower size which is not entirely satisfactory, as intermediates are known, and therefore the course of action taken by Chamberlain seems to offer the best solution to this particular problem. George Forrest was the first to discover the species in 1913, between 3000 and 3300 m in forests on the Shweli-Salween divide in Yunnan near the Burma border. He introduced it into cultivation from that area later in the same year (F 11958). This collection, when selfed, later gave rise to the magnificant clone 'Gargantua'. Forrest subsequently introduced the species again, in 1918 and 1925, and further introductions by other collectors followed. Regrettably, this splendid rhododendron is only suitable for the milder parts of Britain, despite its later flowering season, (June to July) as growth starts too early to escape frosts and it is also liable to bark-split. It is therefore not surprising that the subspecies is less widely grown than one of its hardier offspring — the cultivar 'Polar Bear', a hybrid derived from it. This is a great pity, as subsp. *diaprepes* is one of the loveliest and largest plants in subsection *Fortunea*, fragrant and spectacular when well grown, as can easily be judged from Lilian Snelling's painting; the material from which this was drawn, was provided by Lt.-Col. Stephenson R. Clarke who grew it in his garden at Borde Hill near Hayward's Heath in Sussex, where so many other rhododendrons were successfully cultivated. Under the name *R. diaprepes*, the species received an Award of Merit from the RHS when exhibited in June 1926 by Lionel de Rothschild, Exbury, and a second in 1953 when shown by Mrs R.M. Stevenson, Tower Court, Ascot.

Rhododendron decorum Franchet subsp. **diaprepes** (Balf. fil. & W.W. Sm.) T.L. Ming in Acta Bot. Yunnan. 6(2): 147 (1984).
R. diaprepes Balf. fil. & W.W. Sm. in Notes Roy. Bot. Gard. Edinburgh 10: 100 (1917).
R. rarile Balf. fil. & W.W. Sm., *ibid.* 10: 139 (1917).

Description. An evergreen, spreading to straggly, shrub or tree up to 14 m in height with rough bark; young growth glaucous green. Leaves coriaceous, elliptic-oblong to ovate, 12-19(-30) cm long, 4.4-11(-12.5) cm wide, with rounded base and rounded, minutely mucronate apex, both surfaces of mature leaves glabrous except for persistent punctulate hair-bases on lower surface which is pale glaucous green. Petioles 2-3.5(-5) cm long, glabrous. Inflorescence a lax truss of 5-10 fragrant flowers; rachis 15-20(-30) mm long. Pedicels 15-30(-40) mm long, more or less stipitate-glandular. Calyx rim-like or shallowly to more conspicuously lobed, 2-6 mm long, stipitate-glandular. Corolla fleshy, white or flushed with pink, widely funnel-shaped, 7-8-lobed, (6.5-)8-10.5 cm long, externally more or less glandular, puberulent within. Stamens 18-20, unequal, included, 4-7 cm long; filaments puberulent in lower half. Ovary white stipitate-glandular. Style white stipitate-glandular throughout. Stigma large, discoid, lobulate, 4-5 mm in diameter. Capsule curved, 3-6.5 cm long.

Distribution. Var. *diaprepes* occurs in China (west Yunnan), north-east Burma, and Laos. The range of the type-species is somewhat different, including China (Yunnan, Sichuan and west Guizhou) and north-east Burma, but excluding Laos.

Habitat. Margins of shady, dense forests and more open thickets, 1800-3400 m.

Rhododendron decorum var. *diaprepes*
LILIAN SNELLING

RHODODENDRON FORTUNEI subsp. FORTUNEI

Subgenus *Hymenanthes*
Section *Ponticum* subsection *Fortunea*

'Fortune's rhododendron' is deservedly one of the most popular of the Chinese species. It is fragrant and floriferous, with large flowers which appear late in the season when the danger of frost damage is past, and most of its congeners have ceased flowering. Although it can take up to 15 years for these plants to bloom, once they are mature, they are not only free-flowering, but possess the added advantages of a vigorous constitution and a hardiness rating of 3-4 (on a scale of H1-4, where H1 represents 'tender'). These qualities make *R. fortunei* an admirable subject for a hybridization programme, and since their introduction, both subspecies have been exploited in that direction. As a result, many good hybrids with *R. fortunei* 'blood' in them are readily available to the interested grower. Some of the more familiar offspring of this plant are: 'Albatross' (× *loderi*, white/pink flowers); 'Angelo' (× *griffithianum*, pink/white flowers); 'Ayah' (× *facetum* (*eriogynum*), pink/red flowers); 'Blue Danube' (× 'Purple Splendour', lavender-blue flowers); 'Cadis' (× 'Caroline', pink flowers); 'Golden Belle' (× 'Fabia', yellow flowers); 'Goldsworth Orange' (× *dichroanthum*, salmon-pink flowers); 'James Burchett' (× *catawbiense*, white flowers); 'Sir Frederick Moore' (× 'St Keverne', pink flowers).

Lindley described *R. fortunei* in 1859, naming it after Robert Fortune, a gardener, botanist and plant-collector from Berwickshire, Scotland. Fortune was born on 16 September 1812 and educated at Edrom, after which he became an apprenticed gardener. Later he worked for the elder McNab at the Edinburgh Botanic Garden for two to three years before (in 1842) becoming superintendent of the hothouse department at Chiswick, where the Horticultural Society of London had an arboretum. At this time Fortune was in the employment of the RHS and in 1843 the Society sent him on his first visit to China. He found *R. fortunei* south west of Ning Po in the mountains of the Chekiang province of central China in 1855, but it was not introduced until the autumn of that year, at which time Fortune sent seeds back to Mr Glendinning's nursery at Chiswick. Plants raised from this seed were auctioned in 1859, proving to be the first hardy rhododendron (as distinct from azalea) introduced from China. Six years later, Fortune re-introduced the species and on this occasion, Sunningdale Nurseries were the recipients. Growers were delighted with the new, hardy Chinese rhododendron, and a great deal of interest was aroused.

The accompanying plate was painted by Fitch in 1866 (tab. 5696?) from plants grown from Fortune's seed and supplied by Mr Glendinning from his nursery at Turnham Green, Chiswick. Although the glands on the pedicels, style, etc. are not shown, the illustration is a life-like portrait of this beautiful plant, which is much less widely grown than subsp. *discolor*. This was maintained as a species closely related to *R. fortunei* until 1979 when Chamberlain reduced *R. discolor* to a subspecies of the former (see plate 61) on the basis of leaf shape and the relative leaf width. Thus subsp. *discolor* differs from the type plant in its relatively narrow leaves and tapering leaf bases and it flowers later in the season (July); subsp. *fortunei* has broader leaves with rounded leaf bases and blooms in May.

Rhododendron fortunei is closely related to *R. decorum*, but the latter is a much smaller plant (16 m in height) from much higher altitudes, with smaller leaves and flowers, and pubescent staminal filaments. Both species belong to subsection *Fortunea*, which has affinities with subsections *Grandia*, *Campylocarpa* and *Auriculata*.

Rhododendron fortunei Lindley in Gard. Chron. 1859: 868 (1859), subsp. **fortunei**.
? *R. albicaule* Lév. in Feddes Repert. 18: 148 (1914).

Description. Shrub or tree 2-10 m in height, with an open, rather leggy habit, vigorous and fast-growing when young; leaf-buds long, narrow, tapering, breaking before or with the flower-buds, the bud-scales of both often crimson or pink or stained with red. Leaves coriaceous, dark green above, glaucous beneath, obovate, (8-)18-21 cm long, 2.5-6(-8) cm wide, base more or less tapering, apex acute to rounded, mucronate, veins inconspicuous above, midrib prominent beneath, lower surface reticulate. Petioles reddish, channelled above, 1.2-3 cm long, 2-3(-4) mm in diameter, glabrous. Inflorescence a loose truss of 5-10(-12) large, fragrant flowers; rachis 2-4 cm long, more or less glandular, sometimes glabrous. Pedicels slender, 2.5-3.4(-4.4) cm long, glabrous or stipitate-glandular. Calyx 1-3 mm long, with rounded, more or less distinct lobes, glandular-ciliate or glabrous. Corolla lilac-pink, pale pink or white, or fading to white, open to funnel-campanulate, 4.4-5(-7.5) cm long, with 7 rounded, undulate, more or less crumpled lobes, glandular or glabrous externally, glabrous internally. Stamens 14-16, subequal, somewhat exceeding corolla-tube; filaments flat near base, glabrous; anthers creamy yellow. Ovary usually ovoid-conoid, densely stipitate-glandular, glands yellow-white. Style exceeding stamens, stipitate-glandular throughout, glands yellow-white. Stigma large, capitate. Capsule oblong, ribbed, up to 6 cm long, approximately 12 mm in diameter. Seeds oblong, 4-5 mm long with narrow encircling wing, base and apex rounded, sometimes truncate.

Distribution. China (east Sichuan, Guangxi, Hunan, Guangdong, Jiangxi, Fujian, Anhui, Zhejiang).

Habitat. Woods, forests and along river banks, often also in de-forested areas, 600-2300 m.

Rhododendron fortunei subsp. *fortunei*
WALTER HOOD FITCH

RHODODENDRON FORTUNEI subsp. DISCOLOR

Subgenus *Hymenanthes*
Section *Ponticum* subsection *Fortunea*

The French missionary, Farges, was active in east Sichuan between 1891 and 1894, during which period he found a new rhododendron closely related to *R. fortunei*. However, it was from E.H. Wilson's seed of Farges' plant, sent to Veitch at Coombe Hill Nursery in 1900, that plants were raised, eventually flowering in 1911. Wilson returned to China to collect further seed for the Arnold Arboretum in 1907, and plants from this later consignment were raised at Kew where they flowered in 1917. These plants were considered to be of a new species, which was then described and figured in *Curtis's Botanical Magazine* (tab. 8696) as *R. discolor*. The plate published here, was prepared by Valerie Price for *The Kew Magazine* in 1985-86, from a plant growing at Wakehurst Place, Sussex.

Rhododendron fortunei and *R. discolor* were thought to be two distinct species for many years and it was not until 1979, while engaged on his revision of subgenus *Hymenanthes*, that Chamberlain of the Royal Botanic Garden Edinburgh reduced both species to subspecies of *R. fortunei*. He based his classification on leaf shape and the relative leaf width. Thus subsp. *discolor* differs from subsp. *fortunei* in its relatively narrow leaves and tapering leaf-bases and it flowers later in the season (July); subsp. *fortunei* has broader leaves with rounded leaf-bases and blooms in May. Even today, subsp. *discolor* is the more widely grown subspecies, although both are still extensively used in hybridization programmes.

Rhododendron fortunei Lindley subsp. **discolor** (Franchet) Chamberlain in Notes Roy. Bot. Gard. Edinburgh 37: 330 (1979); 39: 234 (1982).
R. discolor Franchet in J. Bot. (Morot) 9: 391 (1895).
R. mandarinorum Diels in Bot. Jahrb. Syst. 29: 510 (1900).
R. houlstonii Hemsley & Wilson in Kew Bull. 1910: 110 (1910).
R. kirkii Millais, Rhododendrons (ed. 1), 169 (1917) *nom. illeg.*
R. kwangfuense Chun & Fang in Acta Phytotax. Sin. 6: 170, t. 41 (1957).

Description. Shrub or tree 2-10 m in height, with an open, rather leggy habit, vigorous and fast-growing when young; leaf-buds long, narrow, tapering, breaking before or with the flower-buds, the bud-scales of both often crimson or pink or stained with red. Leaves coriaceous, dark green above, glaucous beneath, narrowly to broadly oblanceolate, 8-18(-21) cm long, 2.5-6 cm wide, 2.8-4 times as long as wide; base more or less tapering; apex acute to rounded, mucronate; veins inconspicuous above, midrib prominent beneath, lower surface reticulate. Petioles reddish, 1.2-3 cm long, 2-3(-4) mm in diameter, channelled above, glabrous. Inflorescence a loose truss of 5-10(-12) large, fragrant flowers, rachis 2-4 cm long, more or less glandular, sometimes glabrous. Pedicels 2.5-3.4(-4.4) cm long, slender, glabrous or stipitate-glandular. Calyx 1-3 mm long, with rounded, more or less distinct lobes, glandular-ciliate or glabrous. Corolla lilac-pink, pale pink or white or fading to white, open to funnel-campanulate, 4.4-5(-7.5) cm long, with 7 rounded, more or less crumpled undulate lobes, glandular or glabrous externally, glabrous internally. Stamens 14-16, subequal, somewhat exceeding corolla-tube; filaments flat near base, glabrous; anthers creamy yellow. Ovary usually ovoid-conoid, densely stipitate-glandular, glands yellow-white. Style exceeding stamens, stipitate-glandular throughout, glands yellow-white; stigma large, capitate. Capsule oblong, ribbed, up to 6 cm long, approximately 12 mm in diameter. Seeds oblong, 4-5 mm long with narrow encircling wing; base and apex rounded, sometimes truncate.

Distribution. China (Sichuan, Hubei, Guizhou, Guangxi, Hunan, Anhui and Zhejiang).

Habitat. Open woodlands and thickets, 600-2300 m.

Rhododendron fortunei subsp. *discolor*
VALERIE PRICE

RHODODENDRON AURICULATUM

Subgenus *Hymenanthes*
Section *Ponticum* subsection *Auriculata*

Subsection *Auriculata* is allied to subsection *Fortunea* from which it is distinguished by its setose-glandular shoots and long-cuspidate bud-scales. It contains two species only, *R. chihsinianum*, which is not in cultivation, and *R. auriculatum*, depicted in tab. 8786. Flowering in July and August, *R. auriculatum* stretches the rhododendron season by a significant amount, as well as suggesting itself as a potential parent of a race of late-flowering hybrids, hardy enough to survive and thrive in the English climate. The late flowering period constituted somewhat of an obstacle to would-be hybridists of the species, who found it difficult in July and August to find another compatible rhododendron in bloom! With perseverance the obstacle was surmounted and some fine hybrids have resulted from *R. auriculatum* crosses, one of which, 'Polar Bear' is particularly popular today.

Rhododendron auriculatum is a magnificent species with large leaves and loose trusses of proportionately large, fragrant white or pinkish white flowers. It towers over many other rhododendrons, reaching more than 6 m in height — indeed it can attain 11 m in optimum conditions — and is a glorious plant for a woodland garden, perfuming the air around with its apple-like scent. The species is not common in the wild, but occurs scattered throughout the woods of west Hubei, east Sichuan and east Guizhou. It was discovered and collected near Ichang (west Hubei) by Professor Augustine Henry in 1885. Four years later the species was described by Hemsley, but it was not introduced into cultivation until 1901 when E.H. Wilson, collecting for Veitch, found it near Fang. Subsequently it was found by other collectors in the regions mentioned above.

Rhododendron auriculatum first flowered in England in 1912 at Caerhays Castle, Cornwall, but at Kew, although it had been acquired from Messrs Veitch in 1908, it did not bloom until 1917. It is a surprisingly hardy species, surviving extremes of heat and cold, but needs surrounding vegetation to protect its large leaves from drying winds and hot sun. There is the risk, too, that growth starting so late in the year, the wood may not be ripened enough to withstand early frosts, but that is a risk worth taking. Tab. 8786, reproduced here, was painted by Matilda Smith and published in 1919. *Rhododendron auriculatum* received an Award of Merit from the RHS in 1922 when it was exhibited by Lord Aberconway of Bodnant.

Rhododendron auriculatum Hemsley in J. Linn. Soc. London, Bot. 26: 20 (1889).

Description. A shrub or small tree, rounded or flat-topped, up to 6(-11) m in height with sturdy branches; young growth glandular. Perulae crimson, long-acuminate and ribbon-like. Leaves coriaceous, oblong to oblanceolate, (9.5-)15-20(-32) cm long, (2.8-)4.5-8(-12) cm wide, with auriculate base, rounded apiculate apex and glandular-ciliate margin, pale green and glabrescent above, sparsely villous and glandular beneath, especially on the midrib and veins. Petioles 1-3 cm long, densely stipitate-glandular. Inflorescence a lax, 6-15-flowered truss of fragrant flowers; rachis 20-50 mm long. Pedicels stout, 20-25 (-35) mm long, stipitate-glandular. Calyx about 2 mm long, a minutely lobed rim, the lobes rounded and sparsely glandular. Corolla white, cream or rosy pink, greenish within base of tube, funnel-shaped, 8-11 cm long. Stamens 14, unequal; filaments glabrous. Ovary densely stipitate-glandular. Style glandular throughout. Stigma large, capitate. Capsule purplish, cylindrical, ribbed, 2-3.5 cm long.

Distribution. China (east Sichuan, west Hubei, east Guizhou).

Habitat. Scattered throughout woodland, usually in shade, 500-2300 m.

Rhododendron auriculatum
MATILDA SMITH

RHODODENDRON GRANDE

Subgenus *Hymenanthes*
Section *Ponticum* subsection *Grandia*

Discovered by Griffith in Bhutan in 1847, described by Wight in the same year and introduced by Hooker in 1850 from Sikkim, this handsome species is suitable for only the mildest of the larger gardens in Britain. The mature leaves as well as young growth are susceptible to wind damage and even in mild gardens the species should be grown in a sheltered position.

Rhododendron grande is a spectacular, though variable, plant, with white, silver or fawn undersides to the leaves, pink to crimson bud-scales and flowers which begin life as pink buds and fade to clear white, cream or pale yellow as they age, with a purple blotch at the base of each corolla.

The accompanying plate (tab. 5054) is of the plant originally named *R. argenteum*, the leaves of which have a silvery white indumentum covering their lower surface. Hooker found this form at elevations of 2500-3000 m on 'the summit of Sinchul, Suradah and Tonglo' and thinking it new, described it as *R. argenteum* in 1849 from his own collected specimens. It flowered for the first time in cultivation under glass at Kew in March 1858, when it was drawn by Fitch for *Curtis's Botanical Magazine*. The usual flowering period has since proved to be from February to April. The name *R. argenteum* was reduced to synonymy in 1887.

Fitch's unfinished original painting of this species, from which tab. 5054 was lithographed for publication, admirably demonstrates his working methods.

Subsection *Grandia* is closely related to subsection *Falconera*, differing mainly in its indumentum; (the leaves of subsection *Grandia* have rosulate or dendroid hairs, those of subsection *Falconera* have cup-shaped hairs). Subsection *Grandia* is possibly also related to subsection *Fortunea*, but more distantly so.

In addition to *R. grande*, the subsection embraces 10 species from north-east India, north-east Upper Burma and China, but *R. grande* is the only species in the subsection which is native to the Sikkim Himalaya. Although this species was first introduced by Hooker, his was not the only introduction. R.E. Cooper collected the species in Bhutan in 1914 and Kingdon Ward found it in north-east Upper Burma in 1937. It is usually a denizen of forests and is common at altitudes above 2000 m, where it attains a height of 15 m and develops a trunk 0.75 m in diameter. In cultivation it is somewhat less enthusiastic, rarely exceeding 10 m or so in height. The species was awarded a First Class Certificate in 1901 for a clone with purple-blotched, creamy white flowers, exhibited by F.D. Godman of South Lodge, Horsham, Sussex.

Rhododendron grande Wight in Calcutta J. Nat. Hist. 8: 176 (1847).
R. argenteum Hook. fil., Rhodo. Sikkim Himal. t. 9 (1849).
R. longiflorum Nutt. in Hooker's J. Bot. Kew Gard. Misc. 5: 366 (1853).

Description. A large tree up to 8(-15) m in height, with 1(-3) spreading trunks and rough bark. Leaves coriaceous, elliptic-oblanceolate, 15-27(-46) cm long, 5-9(-13) cm wide, with obtuse to cuneate base and obtuse to acuminate apex; glabrous above; lower surface with a thin, compacted, silvery to fawn indumentum. Petioles 2-3.5(-5) cm long, sparsely floccose, soon becoming glabrous. Inflorescence a 15-25-flowered truss; rachis 4-5 cm long and lengthening in fruit. Pedicels 1.5-3 cm long, densely glandular and sometimes floccose. Calyx a small, undulate rim about 1 mm long, glandular. Corolla fleshy, buds pink opening to pale yellow, cream or blue-white with purple nectar-pouches, oblique, ventricose-campanulate, 8-lobed, 5-7 cm long. Stamens 15-16, unequal, included; filaments glabrous or pubescent at base. Ovary densely stipitate-glandular, or glandular and densely pale brown-tomentose. Style stout, glabrous or sparsely glandular at base. Stigma large, discoid. Capsule 3-4.5 cm long.

Distribution. East Nepal, India (Arunachal Pradesh, Sikkim, Bengal), Bhutan and China (south Xizang).

Habitat. Mixed woodland and on open, stony slopes, also in temperate rain forest, 2100-3200 m.

Rhododendron grande
WALTER HOOD FITCH

RHODODENDRON PROTISTUM var. GIGANTEUM

Subgenus *Hymenanthes*
Section *Ponticum* subsection *Grandia*

Definitely not a plant for the small garden, a specimen of this magnificent giant had to be felled by George Forrest before he could obtain seed. The largest of three trees, it measured almost 24 m in height, well over 2 m in girth 2 m above the ground and was crowned by a canopy approximately 12 m across. A section of the trunk was sent by Forrest to the Royal Botanic Garden at Edinburgh and is preserved in the museum there. *Rhododendron protistum* is closely related to *R. magnificum*. It now comprises two varieties, var. *protistum*, and var. *giganteum* which was previously ranked at specific level. The latter variety is distinguished by the continuous indumentum on the leaf under-surface of mature plants.

Rhododendron giganteum, as it was originally called, was discovered by Forrest, growing near the Yunnan-Burma border, about 80 km north of Teng Chung (Tengyueh) on the N'Maikha-Salween divide in September 1919, and he introduced it into cultivation in the same year (F 18458). At first Forrest assumed that it was rare, but subsequently he and other collectors found the plant in different localities, proving it to be less uncommon than had at first been supposed. Two years later Forrest returned to the area to collect flowering specimens and from these the species was formally described in 1926. He re-introduced the plant in 1924 and 1925, and Kingdon Ward sent seed from Upper Burma in 1926, but *R. protistum* is still rare in cultivation, mainly because it is so tender and only suitable for the mildest areas of Britain. Where it flourishes, var. *giganteum* produces large trusses of pale rose to crimson-purple flowers in February and March, producing a spectacular display.

The form erroneously labelled F 19335 in the Rhododendron Handbook (correctly F 27730) and exhibited from Brodick in the Isle of Arran by the Duchess of Montrose in 1953, was awarded a First Class Certificate by the RHS.

The accompanying plate (tab. 253 new series) was prepared by Ann V. Webster from a flowering branch taken in 1954 from what was, in 1937, the only flowering plant in the country. The plant was growing at Arduaine, near Oban in Argyle and it was Sir Bruce Campbell (of Arduaine) who transported the flowering material to London.

> **Rhododendron protistum** Balf. fil. & Forr. in Notes Roy. Bot. Gard. Edinburgh 12: 131 (1920) var. **giganteum** (Tagg) Chamberlain in Notes Roy. Bot. Gard. Edinburgh 37: 331 (1979).
> *R. giganteum* [Forr. ex] Tagg in Notes Roy. Bot. Gard. Edinburgh 15: 106 (1926).

Description. A large, spreading, evergreen tree, up to 30 m in height, with rough bark and stout, grey-felted young stems. Perulae bright scarlet, caducous. Leaves coriaceous, (12-)20-37(-56) cm long, (4-)8.8-16(-25) cm wide, elliptic or oblanceolate with more or less rounded base and rounded, retuse or minutely apiculate apex. Upper surface glabrous, lower surface with a sparse or discontinuous indumentum (var. *protistum*) or developing a continuous adpressed, buff, woolly indumentum as the plant matures (var. *giganteum*). Petioles stout, 2-3(-5.4) cm long, glabrous. Inflorescence a racemose umbel of 20-25(-30) flowers, sometimes less; rachis stout, 2-3(-4.5) cm long. Pedicels stout, 2-2.5(-3) cm long, densely red-brown or whitish-tomentose. Calyx 2-5 mm long, with 8 irregular, triangular, tomentose teeth. Corolla fleshy, pale rose or white to bright deep rose or crimson-purple with or without speckling, and with basal blotch and dark nectar-pouches, 8-lobed, funnel-campanulate, 5-7.5 cm long. Stamens 16, unequal; filaments 2.5-5.5 cm long, (i.e. both short and long in same flower) glabrous. Ovary densely white or buff-tomentose. Style 4.2-6 cm long, glabrous. Stigma yellow, orange or rarely red, broadly discoid, 6-7 mm in diameter. Capsule 4-5 cm long, tomentose. Chamberlain adds the following note to his description of this species: 'The leaves of young specimens of var. *giganteum* are more or less glabrous beneath and only develop the typical, continuous indumentum as the plants reach maturity. Var. *protistum* may represent an arrested juvenile phase...'

Distribution. China (west Yunnan) and north-east Upper Burma.

Habitat. Mixed woodland, 2450-3350(-4000) m.

Rhododendron protistum var. *giganteum*
ANN WEBSTER

RHODODENDRON MACABEANUM

Subgenus *Hymenanthes*
Section *Ponticum* subsection *Grandia*

Recommended by Kingdon Ward as one of the best of the largest obtainable species of *Rhododendron*, his description of *R. macabeanum* is worth quoting: 'A small scrubby tree with large, broadly oval leaves covered with oyster-grey felt beneath. Flowers in compact dome-shaped trusses, sulphur-yellow with a constellation of purple spots'. A graphic description of a magnificent plant which can be seen at its best in many of the larger gardens in Britain now open to the public. The finest specimens of this long-lived species are in the milder and wetter parts of the country, where the plants tend to reach a greater size more quickly, but the species is surprisingly hardy and it grows and flowers well in many less favourable locations, withstanding winds, drought and even frost, to a greater degree than is to be expected of members of this subsection. Flowers are produced from March to May, which is earlier than other large-leaved species; the flower colour varies in depth, but all shades provide an attractive contrast to the equally attractive foliage. Best of all is the silvery young foliage, gradually emerging from bright crimson perulae which persist as the flowers open.

Rhododendron macabeanum was discovered by Sir George Watt in 1882, while on an expedition to the Naga Hills on the Manipur/Burma frontier, where the species often forms dense, pure stands or grows with birches. A tree of about 15 m in height, it was growing near the summit of Japvo at 2440-2898 m altitude. At the time of its discovery, Watt made copious field-notes as well as preparing a description of the plant, a year later, in 1883, when he named it *R. falconeri* var. *macabeanum* after Mr McCabe, the Deputy Commissioner of Manipur, who had been so helpful to him during his visit. For some unknown reason, the description was not published, until Sir Isaac Bayley Balfour, working from Watt's specimens, described it as a new species, *R. macabeanum*, in 1920 and published both Watt's 1883 manuscript and his own formal description.

The species was introduced into cultivation by Kingdon Ward who visited Japvo and collected seed of *R. macabeanum* in 1927 (KW 7724) and again from a location further to the east, below Mt Sarameti in 1935 (KW 11175). Plants raised from the 1927 seed, flowered 10 to 12 years later, the first to bloom being in Colonel Edward Bolitho's garden at Trengwainton, near Penzance, Cornwall. Colonel Bolitho exhibited a pale-flowered form of the species at the RHS on two occasions and on each occasion the species received an award, the first in 1937 was an Award of Merit, the second in 1938, a First Class Certificate.

The plant depicted by Stella Ross-Craig in the accompanying plate (tab. 187 new series) flowered in the Temperate House at Kew in 1946.

Rhododendron macabeanum [Watt ex] Balf. fil. in Notes Roy. Bot. Gard. Edinburgh 12: 128 (1920).
R. falconeri Hook. fil. var. *macabeanum* Watt, mss.

Description. An evergreen, spreading tree up to 15 m in height in the wild, 9 m in cultivation, (but most plants are not yet fully grown), with rough, light brown bark and stout branches. Leaves clustered at the ends of the branches, coriaceous, broadly ovate to broadly elliptic, 14-25(-31) cm long, 9-18.5(-20) cm wide, with rounded base (cuneate in cultivation) and rounded to retuse, sometimes apiculate apex, upper surface dark green, more or less shiny, glabrous at maturity with impressed reticulate venation, lower surface with a dense greyish white to fawn, more or less woolly, somewhat detersile tomentum. Petioles 2-2.5 cm long, more or less tomentose. Inflorescence a dense, more or less compact, 15-25(-30)-flowered truss. Pedicels stout, 2.5-4 cm long, white-tomentose. Calyx minute, oblique, about 1 mm long, tomentose. Corolla fleshy, pale yellow to lemon-yellow with a purple blotch, ventricose or tubular to narrowly funnel-campanulate, 5(-7.5) cm long, 8-lobed. Stamens 16, unequal, included within corolla; filaments glabrous. Ovary densely reddish-tomentose. Style glabrous. Stigma crimson or pink, large, undulate with central depression. Capsule curved, 2-4 cm long, with more or less persistent tomentum.

Distribution. Apparently confined to, but reasonably common in Manipur and Nagaland in north-east India.

Habitat. It forms dense, often pure stands on the summits of the mountains, sometimes sharing a location with (only) birch, as well as occurring in mixed woodland, 2500-3000 m.

Rhododendron macabeanum
STELLA ROSS-CRAIG

RHODODENDRON HODGSONII

Subgenus *Hymenanthes*
Section *Ponticum* subsection *Falconera*

This is one of the more spectacular plants in the genus *Rhododendron*. It produces large leaves up to 30 cm long, with silver or tawny-felted undersides, followed by large, many-flowered trusses of sometimes fragrant, lilac-pink blooms which fade as they age. Even the bark is an attractive feature, creamy or pink to cinnamon or even mauve in colour and peeling in large pieces. Although hardy in most British gardens, this species is rather too large for many of them, attaining as much as 9 m in height and almost as much in diameter. In the wild, *R. hodgsonii* often dominates, forming impenetrable thickets, not just of twiggy foliage, but also of substantial branches and trunks, in varying situations from valley floors to mountain spurs and rocky slopes, at altitudes between 2900 and 4300 m.

Joseph Hooker reported that, because the wood is so tough and does not split easily, Sherpas and porters in east Nepal use it for making cups, spoons and yak saddles as well as for firewood, while the leaves are employed as plates for customary gifts of butter or curd. Leaves are also used for lining baskets which hold the mashed pulp of *Arisaema* roots.

Presumably tab. 5552, painted by Fitch, was prepared from a plant raised from Hooker's seed, but there is no reference to its source in the accompanying text. It was Griffith who first discovered *R. hodgsonii* in 1838 in Bhutan, but it was not until 1850 that Hooker found the species in Sikkim and introduced it into cultivation. A plant from his original introduction still survives at Stonefield, Argyllshire. The species was re-introduced many times afterwards. At the same time as he introduced *R. hodgsonii*, i.e. 1850, Hooker also introduced *R. falconeri*, a closely related species with which *R. hodgsonii* hybridizes in the wild. Both plants belong to subsection *Falconera*, which is distinguished from the closely allied subsection *Grandia* by its cup-shaped hairs. Apart from the flower colour, a not very dependable distinction, the two species differ in that the ovary and pedicels of *R. hodgsonii* are eglandular, while those of *R. falconeri* are densely viscid-glandular. This difference, together with other supportive characters, adequately separates the two species.

Two forms of *R. hodgsonii* have received awards from the RHS; a clone 'Poets Lawn', exhibited by the Crown Estate Commissioners, Windsor Great Park, received an Award of Merit in 1964, and in 1971 'Harp Wood', exhibited from Sandling Park, was awarded a Preliminary Certificate of Commendation. The species was named after a Mr Hodgson, who was with the East India Company in Nepal at the time the plant was discovered.

Rhododendron hodgsonii Hook. fil., Rhodo. Sikkim Himal., 16, t. 15 (1851).

Description. A large evergreen shrub or small tree 3-11(-12.2) m in height with stout, tomentose branches and peeling bark. Leaves coriaceous, oblanceolate, obovate or elliptic, 17-24(-40) cm long, 6.4-10(-15.6) cm wide with more or less rounded base and rounded to retuse apex, upper surface glabrescent, lower surface with a dense, bistrate silvery to cinnamon tomentum. Petioles 2.5-5 cm long, terete, greyish-floccose. Inflorescence a dense terminal truss of 15-25(-30) flowers; rachis 2.5-5 cm long. Pedicels 2-4 cm long, sparsely greyish floccose. Calyx fleshy, shallowly toothed, 2-3 mm long, sparsely tomentose. Corolla fleshy, pink to magenta or purple with basal blotch, tubular-campanulate, 3-4(-5.8) cm long, 7-8(-10)-lobed. Stamens 15-18, unequal, 1.4-3.4 cm long; filaments glabrous. Ovary tomentose. Style glabrous. Stigma discoid. Capsule 3-4(-6) cm long, narrowly cylindrical, curved, fulvous-tomentose.

Distribution. East Nepal, north India (Arunachal Pradesh, Sikkim, Bengal), Bhutan and China (south Xizang).

Habitat. Open hillsides, in forests, etc., 3000-4000 m, usually at higher altitudes than *R. falconeri*, but the ranges of the two species overlap and hybrids occur where they grow together.

Rhododendron hodgsonii
WALTER HOOD FITCH

RHODODENDRON FALCONERI subsp. FALCONERI

Subgenus *Hymenanthes*
Section *Ponticum* subsection *Falconera*

Joseph Hooker found this plant on Tonglu Mountain in Bengal, growing at an altitude of 3050 m. He described it in 1849, naming it after H. Falconer (1806-1865) who was the Superintendent of Saharanpur Gardens in 1832. Soon afterwards, other collectors found the species in north India, Bhutan and Nepal. Although the species was originally introduced as seed by Colonel Sykes in 1830, it was not until J.D. Hooker re-introduced it in 1850 that the species became widely distributed in cultivation. A very striking plant when not in bloom, due to its large leaves with their tawny or rust-red undersides, *Rhododendron falconeri* is even more spectacular when ornamented with huge trusses of long-lasting, waxy, yellowish white, purple-blotched flowers. Trunks develop as the plant matures, often attaining as much as 60 cm in diameter, and plants can reach a height of 12 m or more. This well-known species is hardy and long-lived, but needs protection from drying winds during the flowering period (April to May). It appreciates plenty of moisture and thrives best along the west coast and other similar mild, wet regions of Britain and elsewhere. Not surprisingly this impressive plant received an Award of Merit from the RHS when exhibited by Messrs Gill of Falmouth in 1922.

Subsection *Falconera* comprises 10 species of which *R. falconeri* is one. In his revision of subgenus *Hymenanthes*, Chamberlain (1982) split *R. falconeri* into two subspecies, subsp. *falconeri*, and subsp. *eximium*, which was previously ranked at specific level; the latter subspecies differs from subsp. *falconeri* in its pink flowers and the scurfy upper surface of its mature leaves, and occurs only in Arunachal Pradesh. In the wild, *R. falconeri* readily crosses with *R. hodgsoni* (see plate 66). Messrs Standish and Noble supplied the material from which tab. 4924 was painted by Fitch in 1856, the year that the species first flowered in Europe; the plant was growing in an open frame at Bagshot, covered only by a mat, not glass, at night. Mr Fairie of Mosely Hill near Liverpool also successfully flowered *R. falconeri* in the same season.

Rhododendron falconeri Hook. fil., Rhodo. Sikkim Himal. t. 10 (1849), subsp. **falconeri**.

Description. A large evergreen tree up to 12 m in height with red-brown, rough, peeling bark; young growth grey-fawn floccose. Leaves coriaceous, broadly elliptic to obovate, 18-35 cm long, 8-17 cm wide, with rounded to cordate base and rounded apex, at maturity upper surface very dark green, rugulose and glabrous, lower surface with a white compacted indumentum overlaid by a dense rufous wool composed of fimbriate cup-shaped hairs. Petioles cylindrical, 2.5-6 cm long, sparsely greyish white-floccose and stipitate-glandular. Inflorescence a dense, terminal, 12-20-flowered truss, 15-22 cm across; rachis stout, 3-6 cm long. Pedicels 4-5.5 cm long, densely stipitate-glandular and sticky. Calyx oblique, about 2 mm long, glandular, sticky. Corolla fleshy, creamy or yellowish white, purple-blotched, obliquely campanulate, 4-5 cm long, 8(-10)-lobed. Stamens 12-16, unequal, included, 1.5-4.5 cm long; filaments usually pubescent at base. Ovary densely glandular, sticky. Style stout, more or less equalling corolla in length, glabrous or glandular at base. Stigma usually yellow, large, discoid. Capsule straight, about 4(-6) cm long, warty with dried-out glands.

Distribution. East Nepal, north India (Arunachal Pradesh, Bengal, Sikkim), Bhutan.

Habitat. Deciduous and mixed forests, 2700-3750 m.

Rhododendron falconeri subsp. *falconeri*
WALTER HOOD FITCH

RHODODENDRON WILLIAMSIANUM

Subgenus *Hymenanthes*
Section *Pontica* subsection *Williamsiana*

A most unlikely candidate for the genus *Rhododendron* to judge by its appearance, this lovely plant with its bronzed young growth, small round leaves and plentiful, outrageously large, nodding pink flower-bells should be in every collection. Space should be no problem, as *R. williamsianum* rarely exceeds 1.5 m in height and is usually smaller. Being a dwarf of dense habit and eventually developing into a dome-shaped bush wider than high, it is an ideal subject for the rock-garden, preferring an open, sunny situation. Frost pockets are lethal to this species, as the young growth is produced too early in the year to escape damage, which can prevent the development of flower-buds. The species is otherwise hardy, easily grown and readily propagated from cuttings or seed and will amply reward the extra trouble of thoughtful siting. It flowers in April and May.

The new subsection *Williamsiana* comprises two species in Chamberlain's 1982 revision (the other is *R. leishanicum*). The subsection appears to be related to subsection *Campylocarpa.*

It is no surprise to discover that *R. williamsianum* has been used most successfully in hybridizing and has left a strong mark on its progeny, most of which have apparently inherited only its good qualities; nor was it entirely unexpected that the species received an Award of Merit from the RHS in 1938 when exhibited by Lord Aberconway of Bodnant.

E.H. Wilson discovered *R. williamsianum* in 1908 on Wa Shan, a mountain in Sichuan, south west of the sacred Emei Shan (Mt Omei), on which it was later found by Chinese botanists. It is very local in occurrence, growing in more or less isolated thickets on cliffs. It has never been found outside Sichuan. Wilson introduced it into cultivation that same year and seed reached England in 1909. Tab. 8935 was prepared by Lilian Snelling from a specimen sent to Kew by J.C. Williams of Caerhays Castle, Cornwall, after whom the plant was named.

Rhododendron williamsianum Rehder & Wilson in Sargent, Pl. Wilsonianae 1: 538 (1915).

Description. An evergreen, spreading, much-branched, dwarf shrub of dense habit up to 1.5 m, but usually less; young growth setose-glandular. Leaves ovate-orbicular, 2-4.5 cm long, 1.4-3.5 cm wide, with cordate base and minutely apiculate rounded apex, glaucous and glabrous on both surfaces except for red sessile glands beneath. Petioles 7-10 mm long, glabrous to setose-glandular. Inflorescence a loose truss of 2-3(-5) nodding flowers. Pedicels 2-3 cm long, glandular. Calyx an undulate or shortly 5-lobed, glandular-ciliate rim about 1 mm long. Corolla pale rose-pink, with or without darker spots, campanulate, 5(-7)-lobed, 3-4 cm long, without nectar-pouches. Stamens 10, about 32 mm long; filaments glabrous. Ovary glandular. Style exceeding stamens, glandular to tip. Stigma small, capitate. Capsule 15-18 mm long.

Distribution. China (central Sichuan).

Habitat. Thickets, on cliffs, 2400-3000 m.

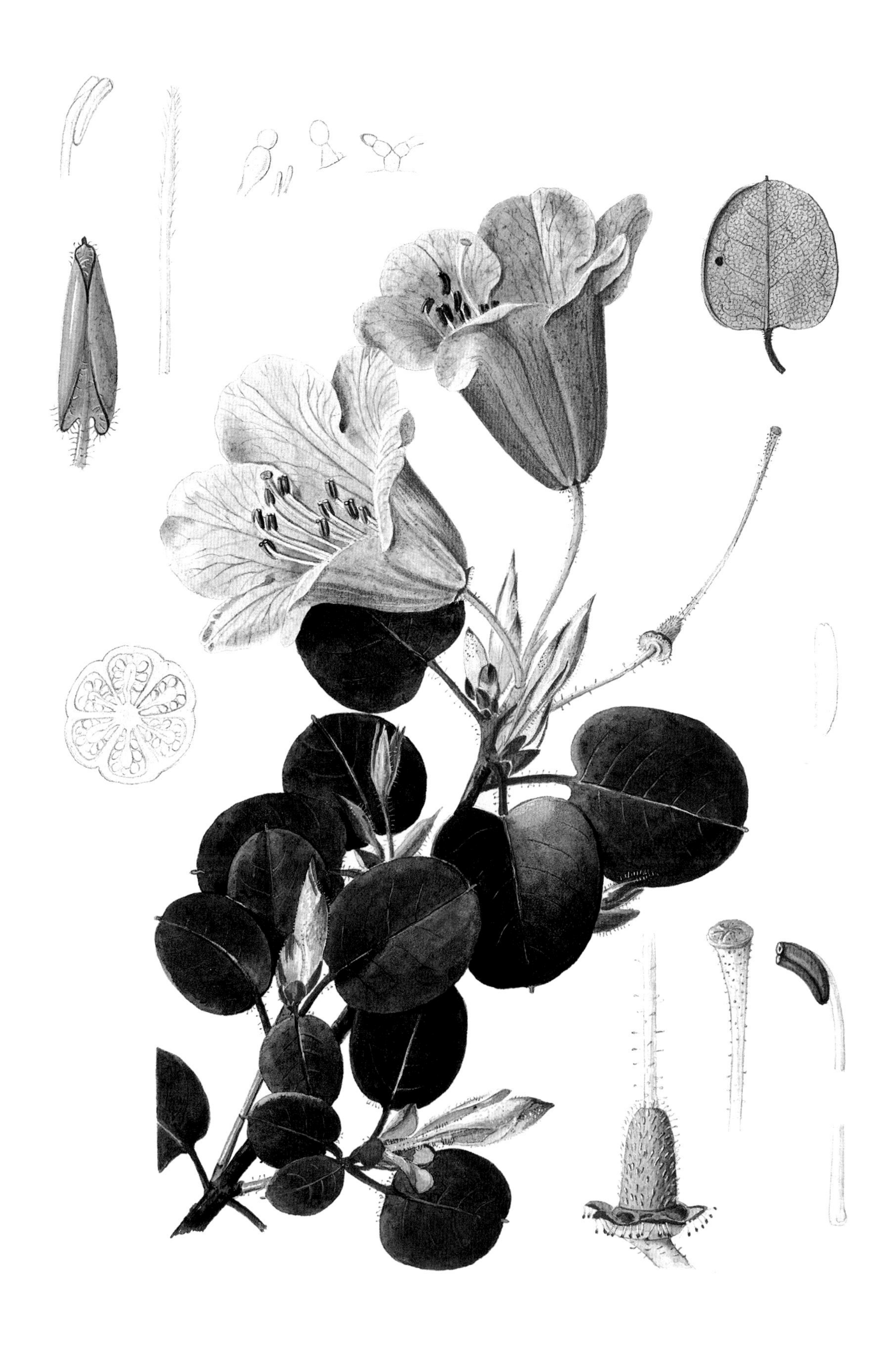

Rhododendron williamsianum
LILIAN SNELLING

RHODODENDRON CAMPYLOCARPUM subsp. CAMPYLOCARPUM

Subgenus *Hymenanthes*
Section *Ponticum* subsection *Campylocarpa*

In Chamberlain's classification, subsection *Campylocarpa* contains four species, *R. callimorphum*, *R. souliei*, *R. wardii* and *R. campylocarpum*. The last-mentioned comprises two subspecies, the type discussed here and subsp. *caloxanthum*, a dwarfer plant which intergrades with subsp. *campylocarpum* and which it tends to replace the further east it grows. Chamberlain does not consider that *R. telopeum* is distinct and has reduced it to synonymy under subsp. *caloxanthum*. In his revision he suggested that subsection *Campylocarpa* is probably more closely allied to subsection *Fortunea*, than to subsection *Thomsonia* as other workers have indicated. One of Hooker's favourite species, *R. campylocarpum* is hardy in most British gardens, although regrettably the flowers of this early-flowering rhododendron are often damaged by frost. If the flower-buds escape such damage, the species makes a fine show in April and May, when it is covered with trusses of translucent pale to lemon-yellow, nodding, bell-shaped, sometimes fragrant corollas. The species usually forms a somewhat aromatic shrub up to 4 m in height which tends to become leggy if afforded too much shade, but can also be tree-like and taller, and for these reasons needs to be carefully sited.

Rhododendron campylocarpum is very closely related to *R. wardii*, but its flowers are campanulate whereas those of *R. wardii* are bowl- or cup-shaped. The main difference between the two species lies in the style, which is glandular throughout its length in *R. wardii* and entirely glabrous, or glabrous for two thirds of its length in *R. campylocarpum*.

The plant was discovered in flower by J.D. Hooker in 1848 in east Nepal, but it was not until November 1849 that he collected seed somewhere between Tumloong and the Cho La on the Sikkim-Xizang frontier. The seed was sent home and produced some compact plants, but the plants commonly found in gardens are of a form with paler flowers and a laxer growth habit than Hooker's original introduction.

The species was figured for *Curtis's Botanical Magazine* by Fitch (tab.4968). A form of *R. campylocarpum* with lemon-yellow flowers, received a First Class Certificate in 1892 when exhibited at an RHS show by Veitch & Sons of Chelsea.

Rhododendron campylocarpum Hook. fil., Rhodo. Sikkim Himal. t.30 (1851) subsp. **campylocarpum**.

Description. A bushy evergreen shrub or small tree up to 4(-6.5) m in height; young shoots usually sparsely stipitate-glandular. Leaves coriaceous, orbicular to oblong or elliptic, 3.2-10 cm long, 1.5-5.4 cm wide with more or less cordate base and rounded, apiculate apex, glabrous when mature, dark glossy green above, often glaucous beneath and occasionally with a few glands at the base. Petioles slender, sometimes stained with red, 0.5-2.4 cm long, usually stipitate-glandular. Inflorescence a 3-10(-15)-flowered raceme; rachis 3-5(-20) mm long. Pedicels 1-3.5 cm long, stipitate-glandular. Calyx (0.5-)3-5 mm long, with more or less unequal, rounded lobes, stipitate-glandular. Corolla pale to sulphur yellow, rarely white, sometimes the buds tinged with red, with or without a faint crimson basal blotch, campanulate, 2.5-4 cm long. Stamens 10, unequal, 1.5-3 cm long; filaments usually sparsely pubescent at base, sometimes glabrous. Ovary densely stipitate-glandular. Style glabrous or glabrous for two-thirds of its length. Stigma wider than style, lobulate. Capsule curved, 1.3-2.5 cm long, stipitate-glandular.

Distribution. The type subspecies is found in Nepal, north India (Arunachal Pradesh, Sikkim), Bhutan and south Xizang.

Habitat. Open forests and stony slopes, 3000-4600 m.

Subsp. *caloxanthum* extends this range eastwards into north-east Upper Burma and adjacent provinces of China (south-east Xizang and west Yunnan). It is found along forest margins, in rhododendron thickets and open places at similar altitudes as the type subspecies, but appears, from the literature, not to occur above 4300 m.

Rhododendron campylocarpum subsp. *campylocarpum*
WALTER HOOD FITCH

RHODODENDRON WARDII var. WARDII

Subgenus *Hymenanthes*
Section *Ponticum* subsection *Campylocarpa*

Of all the yellow-flowered rhododendrons in cultivation, *R. wardii* is arguably the best. 'Best' in many ways, for it is an easy to grow, hardy, compact and free-flowering shrub. It is tolerant of wind, shaded sites, transplanting and pruning, becomes free-flowering at an early age, and has proved to be of great value to hybridists. Its decorative value is widely appreciated, the glaucous leaves providing an admirable foil for the yellow flowers, and the red buds of some forms enhance its appeal. Chamberlain, having removed the species from Tagg's subseries *Souliei* (of series *Thomsonii*) has split *R. wardii* into two varieties (var. *wardii* with yellow flowers and var. *puralbum* with white flowers) and placed it in subsection *Campylocarpa*. George Forrest and Frank Kingdon Ward each collected seeds of *R. wardii* in Yunnan in 1913 at almost the same time. The two collections proved to be of different forms, but both were formally described as *R. wardii* by Smith. Subsequently, as the species became better known, its variability was recognized and the many forms which had been described as new species were relegated to synonymy. As can easily be deduced, *R. wardii* is very common and widespread in the wild, and this fact has resulted in other collectors after Forrest and Kingdon Ward continuing to introduce new forms. Some of those introduced by Ludlow, Sherriff and Taylor are particularly desirable as garden plants, as they produce their flowers towards the end of May and therefore escape all but the latest frosts.

Tab. 587 (new series) is of the type variety and was painted by Margaret Stones in June 1969 from a Ludlow & Sherriff plant grown in Sherriff's garden at Ascreavie, Angus. The plant was raised from seed collected in October 1938 in the Kongbo district of Xizang at an altitude of about 4000 m.

Not surprisingly, this beautiful species has received many awards. The following list is of those awarded by the RHS: AM 1926, exhibited by A.M. Williams, Launceston, Cornwall (as *R. croceum*); AM 1926, exhibited by A.M. Williams, Launceston, Cornwall (as *R. astrocalyx*); AM 1931, exhibited by L. de Rothschild, Exbury, Hampshire; AM 1959, exhibited by Capt. C. Ingram, Benenden, Kent (as the clone 'Ellestree'); AM 1963, exhibited by Crown Estate Commissioners, Windsor, Berkshire (as the clone 'Meadow Pond'); AGM 1969.

Rhododendron wardii W.W. Sm. in Notes Roy. Bot. Gard. Edinburgh 8: 205 (1914) var. **wardii**.
R. mussoti Franchet, *nom. nud.* (Reference not traced).
R. croceum Balf. fil. & W.W. Sm. in Notes Roy. Bot. Gard. Edinburgh 10: 93 (1917).
R. astrocalyx Balf. fil. & Forr. in Notes Roy. Bot. Gard. Edinburgh 13: 30 (1920).
R. prasinocalyx Balf. fil. & Forr., *ibid.* 13: 57 (1920).
R. oresterum Balf. fil. & Forr., *ibid.* 13: 56 (1920).
R. litiense Balf. fil. & Forr., *ibid.* 13: 126 (1920).
R. gloeblastum Balf. fil. & Forr., *ibid.* 13: 200 (1920).

Description. An evergreen shrub or small tree up to 8 m in height with rough, greyish-brown bark and glandular or glabrous, often glaucous young shoots. Leaves oblong or narrowly obovate to broadly ovate, (3-)6-12 cm long, (1.8-)2.3-6.5 cm wide with cordate base and rounded obtuse apex; both surfaces glabrous, glaucous or pale green beneath. Petioles 1-3.5 cm long, glandular or glabrous. Inflorescence a 5-10(-15)-flowered truss; rachis 5 mm or more long. Pedicels 1.5-5 cm long, glandular. Calyx fleshy, 5-15 mm long, usually more or less cupular with unequal, rounded, glandular-ciliate lobes. Corolla white to sulphur-yellow with or without crimson-purple basal blotch, (buds sometimes red or orange) saucer-shaped to cup-shaped, 2.5-4 cm long. Stamens 10; filaments glabrous. Ovary densely glandular. Style glandular throughout. Capsule 2-2.5 cm long, straight or curved.

Distribution. China (west and north-west Yunnan, south-east Xizang).

Habitat. Coniferous or deciduous forests, on open rocky hillsides or by streams, sometimes as undergrowth, sometimes forming impenetrable thickets, 3500-4500 m.

Rhododendron wardii var. *wardii*
MARGARET STONES

RHODODENDRON LONGESQUAMATUM

Subgenus *Hymenanthes*
Section *Ponticum* subsection *Maculifera*

According to Hutchinson, this extremely hardy, slow-growing and free-flowering species was named by Schneider for its shaggy twigs and leaves, amongst which, wrote F.R.S. Balfour of Dawyck, Scotland, the garden warblers frequently build their nests. However, it is much more likely that, as Davidian suggested, Schneider was referring to the persistent long bud-scales which also are a distinctive feature of this species. Material from Balfour's plant, which had been raised from Wilson's seed, was depicted by Lilian Snelling in the accompanying plate, tab. 9430.

The description was based on material discovered and collected by E.H. Wilson (W 3973) in July 1903, at Tatsienlu in west Sichuan; the species was first introduced by him in 1904, and again in 1908 and 1910.

Sometimes *R. longesquamatum* is shy-flowering, but as a rule, after some years, this tendency diminishes and flowers are produced freely in May in cultivation (later in the wild). These cannot be said to be spectacular — indeed, in some forms they are a very grubby or indeterminate shade of pink, but as a garden plant, the species has other, compensating features, its remarkably long, dense branched hairs, pustule-like glands, petaloid calyx and persistent bud-scales, making it a striking variation on the rhododendron theme as well as an interesting foliage plant. The species starts into growth rather early in the year, so despite its hardiness some form of protection from frost is desirable if only as a precautionary measure.

Subsection *Maculifera*, as subseries *Maculiferum*, previously formed part of series *Barbatum*. However, Chamberlain (1982) has suggested that this subsection is probably more closely allied to subsections *Selensia* and *Irrorata*. *Rhododendron longesquamatum* itself has no close relatives.

Rhododendron longesquamatum Schneider, Ill. Handb. Laubholzk. 2: 483 (1909).
R. brettii Hemsley & Wilson in Kew Bull. 1910: 106 (1910).

Description. A compact, evergreen shrub up to 3 m, sometimes 6 m, in height, with rough, shaggy, purplish-brown bark, sturdy densely red-brown shaggy-hairy branches and persistent perulae. Leaves crowded at branch ends, coriaceous, elliptic-oblanceolate, 6-12(-13.5) cm long, 2-4.2 cm wide, with somewhat auricled, subrounded or more or less cuneate base and shortly cuspidate apex, upper surface at first shortly stipitate-glandular and reddish-hairy, the hairs persisting only on the midrib at maturity, lower surface glabrous at maturity except for minute pustule-like glands and the shaggy-hairy midrib. Petioles 1-1.5(-2.6) cm long, densely shaggy with reddish hairs. Inflorescence a terminal, dense, compact 4-6(-12)-flowered truss; rachis up to 2(-8) mm long. Pedicels 15-20(-30) mm long, densely stipitate-glandular. Calyx white or pinkish purple, 'leafy', with 5 unequal, lingulate lobes, 6-14 mm long, stipitate-glandular. Corolla rose-pink with crimson spots and blotch, widely campanulate, 4-4.5 cm long, 5-lobed, lacking nectar-pouches. Stamens 10, unequal, 1.5-3 cm long; filaments hairy towards or at base. Ovary stipitate-glandular. Style stipitate-glandular in lower half. Stigma small. Capsule 1.5-2 cm long, glandular, with persistent calyx-lobes.

Distribution. China (Sichuan).

Habitat. Thickets, woodlands, coniferous forests and grassy slopes, 2300-3500 m.

Rhododendron longesquamatum
LILIAN SNELLING

RHODODENDRON MORII

Subgenus *Hymenanthes*
Section *Ponticum* subsection *Maculifera*

Described in 1911 by Hayata from material collected in Taiwan by U. Mori (for whom it is named) on Randaisan in 1906 and 1908, and by G. Nakahara on Arisan, Mt Morrison in 1906, this rhododendron is endemic to Taiwan, where it is common in mixed forests. Although collected again in 1911, this time by Hayata, the species was not introduced into cultivation until 1918 when E.H. Wilson sent seed back to Britain. Once in cultivation it was re-introduced several times by various collectors. *Rhododendron morii* is a fine, free-flowering, sizeable species which, although hardy, is at its best when given some protection from frost and wind. It produces its large, beautifully marked flowers in April and May. A plant was exhibited by Captain Collingwood Ingram (Benenden, Kent) in 1956, when it received an Award of Merit from the RHS.

Rhododendron morii is closely related to *R. pseudochrysanthum* from which it differs in its larger stature and longer leaves, and with which it hybridizes in the wild. Both species are members of subsection *Maculifera* in Chamberlain's revision (1982), which has affinities with subsections *Selensia* and *Irrorata*.

The accompanying plate was painted by Lilian Snelling in March 1933, from a plant raised at the Royal Botanic Garden, Edinburgh from Wilson's seed (no 10955). The seed was collected in 1918 on the summit of Arisan; it did not come directly from Wilson, but was obtained from Mr J.C. Williams of Caerhays who sent it to Edinburgh in 1919.

Rhododendron morii Hayata in J. Coll. Sci. Imp. Univ. Tokyo 30: 173 (1911).

Description. A large evergreen shrub or small tree up to 8(-10) m in height with rough bark, the branches densely blackish-floccose when young, soon glabrous. Leaves coriaceous, lanceolate-elliptic, (7-)9-15 cm long, 2.8-3.5(-4.1) cm wide, rounded base and acuminate apex, smooth and glabrous above and beneath except for lower surface of midrib which is floccose. Petioles 1.5-2(-3) cm long, glandular and hirsute. Inflorescence a lax, terminal truss of 5-12(-15) flowers; rachis 1-2(-3.6) cm long. Pedicels 2.5-4 cm long, shortly stipitate-glandular. Calyx about 2 mm long, shallowly and unequally lobed, lobes broad, rounded, glandular-ciliate. Corolla white or tinged with pink, usually with red basal blotch and flecks, widely campanulate, 3-5 cm long, lacking nectar-pouches. Stamens 10, unequal, 1.5-3.5 cm long; filaments minutely pubescent at base. Ovary densely tomentose and sparsely stipitate-glandular. Style tomentose at base, otherwise glabrous. Stigma lobulate. Capsule 1-2.5 cm long.

Distribution. Endemic to Taiwan.

Habitat. Mixed forests above 1650 and up to 3500 m.

Rhododendron morii
LILIAN SNELLING

RHODODENDRON SELENSE subsp. DASYCLADUM

Subgenus *Hymenanthes*
Section *Ponticum* subsection *Selensia*

To show to advantage, plants of *R. selense* are best grown in groups, as in many forms, the flower display is modest and the inflorescences are often concealed by the new foliage, which overtops them. The species is hardy, but rarely seen in cultivation. It is many years before *R. selense* reaches the flowering stage, by which time the plant has usually become leggy, especially so when grown singly as a specimen plant. It is a different story in the wild, where the species has excited comment. On its native Chinese mountainsides it is much more free-flowering and covers large areas with a pink haze when the plants are in full bloom in April and May. *Rhododendron selense* has never appeared in either *Curtis's Botanical Magazine* or *The Kew Magazine* and the accompanying plate of subsp. *dasycladum* was specially painted by Mary Mendum (Mary Bates) in Edinburgh for this book, from material growing in the Royal Botanic Garden. Subsp. *dasycladum* may be distinguished from subsp. *selense* by its long gland-tipped hairs or bristles on the stems and petioles. The more distinctive (and more attractive) forms tend to have pinker and larger flowers than does subsp. *selense*.

Soulié, the French missionary, discovered the species in 1895 on the Se La, Mekong-Salween divide and George Forrest introduced subsp. *dasycladum* from the mountains north-west of the Yangtze Bend in 1917. Subsection *Selensia* Sleumer in Chamberlain's revision (1982) includes subseries *Selense* Tagg and subseries *Martinianum* Tagg, both previously placed in series *Thomsonii* Tagg. Chamberlain split the species into four subspecies which are only partially differentiated geographically, three of the subspecies (subspp. *selense*, *dasycladum* and *setiferum*) intergrading where their ranges overlap. The fourth subspecies (subsp. *jucundum*) is isolated geographically but exhibits only small diagnostic characters. The taxonomy of the species is complicated by the suspected existence of hybrid swarms in the wild.

Rhododendron selense Franchet in J. Bot. (Morot) 12: 257 (1898) subsp. **dasycladum** (Balf. fil. & W.W. Sm.) Chamberlain in Notes Roy. Bot. Gard. Edinburgh 39(2): 279 (1982).

Description. An evergreen shrub or small tree up to (2.7-)5 m in height, at first compact, becoming leggy, with rough grey bark and long stipitate- to densely setose-glandular young growth. Leaves more or less herbaceous, ovate, obovate or elliptic, 4-9 cm long, 2.2-4.2 cm wide, with rounded base and rounded, apiculate apex, both surfaces glabrous except for a few persistent hairs beneath near the base. Petioles 1-1.5(-3) cm long, indumentum like that of young growth. Inflorescence a lax cluster of 3-8 flowers; rachis 2-4 mm long. Pedicels 1.5-2(-3) cm long, glandular and with a few hairs. Calyx 2-3 (-5) mm long, lobes rounded-ovate to lingulate, glandular and hairy. Corolla white, pale cream or pink to deep pink, purple-spotted or unmarked, with or without a purple basal blotch, funnel-campanulate, 2.2-4 cm long. Stamens 10, unequal; filaments pubescent at base. Ovary densely glandular. Style glabrous or glandular at base, rarely glandular for half its length. Capsule narrow, curved, 1.2-3.5 cm long.

Distribution (of the species as a whole). China (north-west & mid-west Yunnan, south-west Sichuan and south-east Xizang).

Habitat. A variety of situatuions including stony slopes, open pine forest and rhododendron thickets, 2700-4400 m. Subsp. *dasycladum* replaces subsp. *selense* in the southern part of the range of the species.

Rhododendron selense subsp. *dasycladum*
MARY MENDUM

RHODODENDRON HABROTRICHUM

Subgenus *Hymenanthes*
Section *Ponticum* subsection *Glischra*

This lovely, free-flowering species is closely related to *R. glischrum* subsp. *glischrum*, but is just sufficiently distinct to merit specific status. As a garden plant, it is more attractive than *R. glischrum*, with neat, shorter, broader leaves and produces compact inflorescences of more (and prettier) pink, blotched and finely spotted flowers in April and May. The branches, leaves, pedicels, calyx are all densely clothed with distinctive long, red bristles, to which the specific epithet refers, and it is this feature which attracts the most attention.

Subsection *Glischra* to which this species belongs, is closely related to subsection *Taliensia*. It comprises Tagg's subseries *Glischrum* of series *Barbatum* and also part of Sleumer's subsection *Barbata*. The plant was discovered in Yunnan and introduced from there by George Forrest in 1912, when he found it growing on the western flank of the Shweli-Salween divide. He re-introduced it from the area several times subsequently, as did other collectors, among them Farrer, who collected seed of the species in Upper Burma in 1919, Rock, who re-introduced it from Yunnan in 1922 and Kingdon Ward who found it in north Burma in 1938.

The species flowered at Kew in 1920, but apparently was not figured for *Curtis's Botanical Magazine*. The plate published here was painted by Lilian Snelling in April 1933, from a plant growing in Lord Headfort's garden at Kells, Co. Meath, Eire. In the same year, a plant with pink flowers received an Award of Merit when exhibited at the RHS by H. White of Sunningdale Nurseries.

Rhododendron habrotrichum Balf. fil. & W.W. Sm. in Notes Roy. Bot. Gard. Edinburgh 9: 232 (1926).

Description. An evergreen, somewhat rounded shrub up to 4 m in height with rough bark; young growth densely red-purple glandular-setose. Leaves subcoriaceous, elliptic or ovate to obovate, 7-16(-18) cm long, 3-7.5(-8) cm wide, with rounded base, acute to shortly acuminate apex and ciliate-setulose margins, upper surface very dark green, somewhat rugulose, glabrous, lower surface glabrous except for glandular-setose midrib and main veins. Petioles stout, 1-2(-3) cm long, densely glandular-setose. Inflorescence a usually compact 10-20-flowered truss; rachis about 1 cm long. Pedicels 2-2.5 cm long, densely glandular-setose. Calyx red, deeply divided into 5 oblong, rounded, stipitate-glandular, ciliate lobes 1-1.5 cm long. Corolla white, flushed with pink or pale to deep pink, with crimson-purple basal blotch and fine spots, campanulate, 4-5 cm long. Stamens 10, unequal, 1.5-4 cm long; filaments puberulous at base. Ovary densely glandular-setose. Style glandular-setose in lower third. Capsule about 2 cm long.

Distribution. China (west Yunnan) and north-east Upper Burma.

Habitat. Thickets, rocky slopes and other open situations, 2700-3350 m.

Rhododendron habrotrichum
LILIAN SNELLING

RHODODENDRON CRINIGERUM var. CRINIGERUM

Subgenus *Hymenanthes*
Section *Ponticum* subsection *Glischra*

Seedling plants presented to Kew by Sir John Ramsden furnished the material from which the accompanying plate was painted. The plants were raised from seeds collected by Joseph Rock (no 59186) on Mt Kenichunpu in the north-east Upper Burma/south-east Xizang borderlands of the Irrawaddy-Salween divide, although this was not the first introduction of the species. *Rhododendron crinigerum* was first discovered by Soulié in 1895, growing on the Mekong-Salween divide. It was not introduced into cultivation until 1914, when George Forrest found it in the same general area, after which it was re-introduced several times, proving itself to be more widely distributed than had been thought at altitudes ranging between 3050 and 4423 m.

It was Forrest who discovered *R. crinigerum* var. *euadenium* in July 1924, on the Salween-Kiu Chang divide in north-west Yunnan. This variety occurs at altitudes between 3355 and 4300 m on cliffs and rocky slopes, in forests and along their margins, over the range of the type species. It differs from var. *crinigerum* in the leaf under-surface which is more densely glandular and has a sparse indumentum, the latter character contrasting with the dense, matted indumentum of the type variety; var. *euadenium* is also a little less hardy than the type. Flowers of both varieties appear in April and May.

Within the subsection *Glischra*, *R. crinigerum* is most closely related to *R. recurvoides* which is a much smaller shrub (classed as 'dwarf' by Chamberlain) with smaller, blunt leaves.

As can be judged from Lilian Snelling's plate (tab. 9464) *R. crinigerum* is, in its best forms, a handsome rhododendron, and one of these with pale pink, heavily red-blotched and spotted flowers, not surprisingly received an Award of Merit from the RHS when shown from Exbury in 1935.

Rhododendron crinigerum Franchet, J. Bot. (Morot) 12: 200 (1908) var. **crinigerum**.
R. ixeunticum Balf. fil. & W.W. Sm., Notes Roy. Bot. Gard. Edinburgh 9: 240 (1916).

Description. An evergreen shrub or small tree up to 5(-6) m in height, with sparsely stipitate-glandular (sticky) young growth. Leaf buds sticky. Leaves subcoriaceous, obovate-oblanceolate, (7-)10-17(-20) cm long, (2.3-)3-4.2(-5) cm wide with rounded base and cuspidate apex; upper surface glabrous, sometimes bullate, lower surface with (usually) a dense, matted, fawn to cinnamon tomentum and stipitate glands. Petiole 1-2 cm long, densely stipitate-glandular. Inflorescence a truss of 8-14(-20) flowers; rachis 1-1.5 cm long. Floral bud-scales persistent during flowering. Pedicels 2.5-3(-4) cm long, densely stipitate-glandular and sparsely hairy. Calyx 5-10 mm long, densely stipitate-glandular; lobes unequal, oblong or ovate, rounded. Corolla white flushed pink, more or less speckled and with a basal blotch, campanulate, 3-4 cm long. Stamens 10, unequal, 1-3 cm long, declinate; filaments pubescent in lower part. Ovary stipitate-glandular with glabrous, lobed, basal disc. Style exceeding stamens, glabrous or glandular at base. Stigma lobulate. Capsule about 1.5 cm long, densely glandular.

Distribution. China (widespread in north-west Yunnan and south-east Xizang) north-east Upper Burma.

Habitat. Open pine forests or along their margins and on rocky slopes or cliffs, 3350- 4300 m.

Rhododendron crinigerum var. *crinigerum*
LILIAN SNELLING

RHODODENDRON VENATOR

Subgenus *Hymenanthes*
Section *Ponticum* subsection *Venatora*

Until comparatively recently classified as a member of series *Irroratum* subseries *Parishii*, but with no close allies in that or any other group, *R. venator* was transferred by Chamberlain in 1979, to the monotypic subsection *Venatora*. This subsection, he suggested, has probable links with subsection *Maculifera* as well as with subsection *Irrorata*. The specific epithet *venator*, meaning hunter, presumably refers to the scarlet flowers, which in colour approach that of the jackets worn by the hunting fraternity. Frank Kingdon Ward discovered the plant late in 1924, in the Tsangpo Gorge in south-east Xizang. It was abundant in swampy places everywhere, but especially around Pemakochung. The species proved to be hardy and late-flowering (May to June), thereby missing the frosts, so that by 1933 it was not only established, but also flowering in some British gardens. Tagg described the species in 1934, since when it has been used in hybridizing with considerable success. The species was re-introduced in 1946/47 by Ludlow, Sherriff and Elliott from Pemakochung to which region the species appears to be confined.

The accompanying plate was specially painted by Mary Mendum (Mary Bates) for this book, at the Royal Botanic Garden, Edinburgh in 1989, because the species had not been featured in either *Curtis's Botanical Magazine* or *The Kew Magazine*.

A plant with reddish orange flowers, raised from Kingdon Ward's no. 6285, was exhibited by the Hon. H.D. McLaren, Bodnant, in 1933 at the RHS when it received an Award of Merit.

Rhododendron venator Tagg in Notes Roy. Bot. Gard. Edinburgh 18: 219 (1934).

Description. A straggling, evergreen shrub up to 3 m in height in cultivation (much less in the wild) with greyish brown bark; young growth setose-glandular and floccose. Leaves elliptic-lanceolate, 8.5-15 cm long, 2-2.4(-4.5) cm wide, with rounded base and acute to acuminate apex, both surfaces green and glabrous except for a thin, stellate and folioliferous indumentum on the midrib beneath. Petioles stout, 1-1.5 cm long, stellate-tomentose and setose-glandular. Inflorescence a compact truss of 7-10 flowers; rachis about 5 mm long. Pedicels about 1 cm long, reddish stellate-hairy and glandular. Calyx 3-5 mm long, tomentose and glandular at base with rounded glandular-ciliate lobes. Corolla fleshy, reddish orange, scarlet or crimson with darker nectar-pouches, tubular-campanulate, 3-3.5(-3.8) cm long, 5-lobed. Stamens 10; filaments glabrous. Ovary densely tomentose and glandular. Style glabrous, or hairy at base. Capsule about 2 cm long, curved.

Distribution. China (south-east Xizang).

Habitat. Swamps and on rock faces, 2400-2600 m.

Rhododendron venator
MARY MENDUM

RHODODENDRON ARAIOPHYLLUM

Subgenus *Hymenanthes*
Section *Ponticum* subsection *Irrorata*

One of the most attractive species in subsection *Irrorata*, this species was previously placed in series *Irroratum* subseries *Irroratum*. A graceful plant, *R. araiophyllum* is eglandular and of slender growth habit, with red young growth and narrow, long-pointed leaves. The specific epithet refers to the relatively thin texture of the leaves. The cup- or bowl-shaped flowers are pure white or white flushed with pink, with a bright crimson blotch in the throat breaking into speckles on the upper lobes. These features combine to make *R. araiophyllum* a highly desirable plant for the garden, but the species is tender and produces growth early in the year, flowering in April and early May, which precludes it from being grown in all but the very mildest locations. The species most closely related to *R. araiophyllum* is *R. annae*, from which the former species is distinguished by its glabrous style and lack of punctate hair-bases. The subsection *Irrorata* is closely allied to subsections *Maculifera* and *Venatora*, differing from both in the character of its indumentum.

First discovered by George Forrest on the Shweli-Salween divide in 1913, and introduced by him in 1917 from the same area, this species was re-introduced by Forrest many times subsequently. Seed was collected by Farrer in 1919 in north-east Upper Burma and by Kingdon Ward in north-west Yunnan in 1922. Davidian recorded (1989) that, in all, this charming species has been introduced on 18 occasions, yet it is not often grown, perhaps because it is so often spoiled by frost. The accompanying plate by Lilian Snelling was never published, and is reproduced here, being both an accurate representation and one which conveys the charm of the plant. The species received an Award of Merit in 1971, when a clone named 'George Taylor', with white flowers, blotched and spotted with crimson and grown at Wakehurst Place, Sussex, was exhibited at the RHS.

Rhododendron araiophyllum Balf. fil. & W.W. Sm. in Trans. Bot. Soc. Edinburgh 27: 189 (1917).

Description. An eglandular, evergreen shrub or small tree up to 6.5 m in height with slender branches; young growth usually brightly coloured, floccose, soon glabrous. Leaves lanceolate or narrowly elliptic, 5.5-13 cm long, 1.8-3.2 cm wide, with cuneate base, acuminate apex and flat to undulate margin, both surfaces glabrous at maturity, except for the midrib beneath, which sometimes remains floccose. Petioles 7-15 mm long. Inflorescence a lax, 5-10-flowered raceme; rachis 5-15 mm long. Pedicels slender, about 1.5(-2.5) cm long, and minutely hairy. Calyx a minute, undulate rim up to 2 mm long, glabrous or sparsely ciliate. Corolla white, or white flushed with pink, with crimson blotch and spots, or the spots lacking, open-campanulate, 5-lobed, 2.8-3.5(-4) cm long. Stamens 10, unequal, 1.4-2.8 cm long; filaments pubescent in lower part. Ovary with sparse, short, white hairs. Style often red, slender, glabrous. Stigma lobulate. Capsule 1-1.6(-2.2) cm long.

Distribution. China (west central Yunnan) and north-east Upper Burma.

Habitat. Mixed forest and open scrub, 2300-2400(-3350) m.

Rhododendron araiophyllum
LILIAN SNELLING

RHODODENDRON KENDRICKII

Subgenus *Hymenanthes*
Section *Ponticum* subsection *Irrorata*

Although it was Griffith who originally discovered this plant in Bhutan, the species was first described by Nuttall in 1853, from a specimen collected by Nuttall's nephew, Thomas Booth. Booth's plant also was growing in the mountains of Bhutan, at about 2135 m and it is thought that the seed he collected was the first to be introduced, but there is no way of confirming this fact. What is certain is that Cooper re-introduced the species in 1915; Balfour then published an expanded description from Cooper's material in 1917 and Kingdon Ward sent seed back from Assam in 1935 and 1938. *Rhododendron kendrickii* was named after a friend of Nuttall's, Dr Kendrick MD, who, for a time, was President of the Warrington Natural History Society. It is a tender species, starting into growth early in the season and, like so many rhododendrons, is therefore liable to frost damage. It has relatively narrow leaves and, belying the accompanying plate, flowers are reputed to be small; they are a glowing deep rose, scarlet or crimson, with a variable amount of spotting in the throat, and are produced in April and May. Although not common in cultivation, the species is grown successfully in some of the milder parts of southern England. The subject of Fitch's painting (tab. 5129) is named var. *latifolium*. Chamberlain (1982) however, cites the plate as *R. kendrickii*. No mention of var.*latifolium* is made in the synonymy. It appears to be merely a broad-leaved variant of *R. kendrickii* which does not merit taxonomic status.

Rhododendron kendrickii Nutt. in J. Bot. 5: 358 (1853).
? *R. shepherdii* Nutt., *ibid.* 5: 360 (1853).
R. pankimense Cowan & K. Ward in Notes Roy. Bot. Gard. Edinburgh 19: 180 (1936).

Description. A broadly erect, evergreen shrub or small tree up to 8 m in height with light brown bark and slender branches. Leaves more or less coriaceous, narrowly elliptic-oblanceolate, 10-13.5(-17) cm long, 2-3.5 cm wide, with cuneate to rounded base and acuminate apex, margin wavy, upper surface glabrous at maturity, lower surface glabrous, except for the hairy midrib, persistent punctate hair-bases lacking. Petioles 1-1.5 cm long, glabrescent. Inflorescence a dense, terminal, 10-20-flowered truss; rachis up to 15 mm long. Pedicels 4-8 mm long, tomentose. Calyx 2-3 mm long, lobes rounded, glabrous or glandular-ciliate. Corolla fleshy, deep rose to scarlet or crimson, more or less spotted with blackish crimson, tubular-campanulate with basal nectar-pouches, 3-4 cm long. Stamens 10, unequal, 1.2-2.2 cm long; filaments glabrous. Ovary eglandular but with some reddish branched hairs. Style glabrous. Capsule 1.4-3 cm long.

Distribution. North-east India (Arunachal Pradesh), Bhutan and China (south Xizang).

Habitat. Coniferous and mixed forest, (2135-)2300-2800 m.

Rhododendron kendrickii
WALTER HOOD FITCH

RHODODENDRON ANTHOSPHAERUM

Subgenus *Hymenanthes*
Section *Ponticum* subsection *Irrorata*

Diels described this species from a plant collected by George Forrest in 1906 below the Sung Kwei pass in north-west Yunnan, but some years earlier, in 1887, a plant was collected by Delavay at Mosoyu in the Tali valley, and labelled *R. irroratum*. This was subsequently found to be referable to Diels' new species. Later gatherings by various collectors proved the species to be both variable and widely distributed in west Yunnan, south-east Xizang and north-east Upper Burma.

Rhododendron anthosphaerum was introduced into cultivation by Forrest in 1910 (as F 5848) and re-introduced by him many times afterwards, as well as by numerous other collectors. It appears to be common throughout its range and colour forms abound, but it is not widely cultivated, and authors in general are lukewarm regarding its charm. The species produces plentiful flowers from March to May and ample fertile seed thereafter, but it is not entirely hardy, needing some protection in colder gardens. Even in milder areas frost damage is inevitable, though not necessarily irreparable, because the plant starts into growth so early in the year, i.e. in February and March. Subsection *Irrorata* comprises 18 species and embraces Tagg's series *Irroratum* subseries *Irroratum*. In Chamberlain's revision (1982), the synonymy of the species includes *R. gymnogynum*, *R. chawchiense* and *R. eritimum*, but with reservations as to their geographical differentiation. The plant from which Lilian Snelling painted tab. 9083 was raised from Forrest's seed (F 5848) by E.J.P. Magor of Lamellen, St Tudy, Cornwall.

Rhododendron anthosphaerum Diels in Notes Roy. Bot. Gard. Edinburgh 5: 215 (1912).
R. eritimum Balf. fil. & W.W. Sm. in Trans. Bot. Soc. Edinburgh 27: 190 (1917).
R. hylothreptum Balf. fil. & W.W. Sm., *ibid.* 27: 195 (1917).
R. gymnogynum Balf. fil. & Forr. in Notes Roy. Bot. Gard. Edinburgh 13: 47 (1920).
R. heptamerum Balf. fil. in Notes Roy. Bot. Gard. Edinburgh 13: 48 (1920).
R. chawchiense Balf. fil. & Farrer in Notes Roy. Bot. Gard. Edinburgh 13: 247 (1922).
R. persicinum Hand.-Mazz. in Akad. Wiss. Wien, Math.-Naturwiss. Kl., Denkschr., Anz. 60: 97 (1924).

Description. An evergreen shrub or small tree up to 12 m in height with grey to brown bark; young growth rufous-hairy and stipitate-glandular. Leaves thinly coriaceous, elliptic-obovate to oblong, 6-16(-18.5) cm long, 2-4.5(-5) cm wide, with a more or less cuneate base, acute to acuminate apex and more or less flat margin, upper surface glabrous at maturity, lower surface with some persistent red punctate hair-bases on the veins. Petioles 1-2 cm long, glabrescent. Inflorescence a more or less dense, terminal, 10-15-flowered truss; rachis 5-10 mm long. Pedicels 10-15(-20) mm long, rufous-hairy. Calyx oblique, cupular, minutely 5-7-lobed, 1-2 mm long, glabrous or stipitate-glandular. Corolla pale pink to rose-magenta or crimson, lilac to magenta or pale peach, occasionally white, with or without crimson speckles and with or without a basal blotch, tubular-campanulate with nectar-pouches, 3-4.5(-5.9) cm long, 6-7-lobed. Stamens 10-14, unequal, 1.2-4.2 cm long; filaments glabrous or puberulous at base. Ovary usually glabrous, occasionally sparsely rufous-hairy. Style glabrous. Stigma small, lobulate. Capsule 2-2.5(-3) cm long.

Distribution. China (Yunnan, south-east Xizang), north-east Upper Burma.

Habitat. Open rocky slopes or in thickets and various kinds of forest, 2700-4000 m.

Rhododendron anthosphaerum
LILIAN SNELLING

RHODODENDRON YAKUSHIMANUM subsp. YAKUSHIMANUM

Subgenus *Hymenanthes*
Section *Ponticum* subsection *Pontica*

This hardy species needs no introduction, being one of the most popular, most enthusiastically recommended and most written-about rhododendrons to enter the limelight in recent years. Parent of innumerable new hybrids, including those named after the 'Seven Dwarfs' (i.e. 'Bashful', 'Doc', 'Dopey', 'Grumpy', 'Happy', 'Sleepy', 'Sneezy') and unsurpassed for container growing, there must be scarcely a garden in Britain in which the species is not represented by one form or another. As its name suggests, *R. yakushimanum* grows wild only on the Japanese island of Yakushima, where it is protected by the Japanese government, a necessary precaution if the species is to survive in the wild.

This rhododendron is a slow-growing, compact, free-flowering plant which will tolerate full exposure to both sun and wind. The leaves are small, the flowers large for the size of the plant and the indumentum of the lower leaf-surface is another attractive feature. A further point in its favour is that it flowers later in the season, towards the end of May when many rhododendrons have more or less finished blooming. Propagation can readily be accomplished by seed or cuttings, and the resulting plants will flower at an early age.

Rhododendron yakushimanum was described by Nakai in 1921 but was first introduced into cultivation in 1934 when Lionel de Rothschild received two plants sent to him at Exbury from K. Wada's Nurseries, Numazushi, in Japan. One of them was grown on at Wisley, from where the clone named 'Koichiro Wada' was exhibited at the Chelsea Show in 1947 and was awarded a First Class Certificate. Subsequent introductions of wild seed have produced plants showing wide variation, and it is obvious that the full potential of the species as a parent has not yet been realized, so that although much work with the species and its progeny has been accomplished, much remains to be done. The plant in its diverse forms is becoming more and more readily available commercially and correspondingly, the popularity of the species continues to increase. As yet the demand has not been sated.

Subsection *Pontica* Sleumer includes series *Ponticum* sensu Tagg. According to Chamberlain (1982) the subsection is probably related to subsection *Argyrophylla*. In his 1982 classification, Chamberlain reduced *R. makinoi* to a subspecies of *R. yakushimanum* as follows:

Leaves 2.3-6 times as long as broad; perulae deciduous · 4a. subsp. *yakushimanum*
Leaves 7.5-10 times as long as broad; perulae persistent · 4b. subsp. *makinoi*

More recently Hara, Chamberlain and Doleshy have resurrected the two as species. *Rhododendron yakushimanum* is illustrated in the accompanying plate which was painted by Margaret Stones in 1976, from a plant donated to Kew by Mr Geoffrey Gorer, which had been growing at Wakehurst Place, Sussex, since 1970. The plant was raised from seed collected by Mr Frank Doleshy of Seattle on the mountain of Kuromi dake in October 1965 and distributed through the American Rhododendron Society. This is not the famous 'Koichiro Wada' clone mentioned above, but a more recent introduction.

Rhododendron yakushimanum Nakai in Bot. Mag. (Tokyo) 35: 135 (1921). subsp. **yakushimanum**.
R. metternichii Sieb. & Zucc. var. *yakushimanum* (Nakai) Ohwi in Bull. Natl. Sci. Mus. 33: 81 (1953).
R. metternichii Sieb. & Zucc. subsp. *yakushimanum* (Nakai) Sugimoto, New Key Jap. Trees, 470 (1961).
? *R. metternichii* Sieb. & Zucc. var. *intermedium* Sugimoto, *ibid.* 471 (1961).
R. degronianum Carrière var. *yakushimanum* (Nakai) Kitamura in Acta Phytotax. Geobot. 25: 38 (1972).

Description. A dense, compact to open, evergreen shrub up to 2.5 m in height (up to 1 m in cultivation) with rough bark; young growth floccose. Leaves coriaceous, narrowly to broadly elliptic or linear-lanceolate, 6-21 cm long, 1-3(-5) cm wide, with rounded to cuneate base and rounded to acute apex, upper surface glabrous or thinly floccose near base, lower surface thickly white- to fulvous-tomentose. Petioles 1-1.5 cm long, tomentose, soon glabrescent. Inflorescence a lax truss of 5-10 flowers; rachis 2-5 mm long. Pedicels 15-25 mm long, densely fulvous-tomentose. Calyx 2-5(-7) mm long, tomentose. Corolla deep pink opening to pale rose and fading to white, with or without deeper flecks, funnel-campanulate, 3-4 cm long. Stamens 10, unequal; filaments glabrous. Ovary densely whitish- to brown-tomentose. Style glabrous. Capsule about 15 mm long, sometimes more.

Distribution. Subsp. *yakushimanum* is endemic to the island of Yakushima. (*R. makinoi* is confined to forests in central Honshu at the lower altitudes of 180-700 m. It flowers in June).

Habitat. Conifer forest and exposed mountain tops among rocks, 1200-1800 m.

Rhododendron yakushimanum subsp. *yakushimanum*
MARGARET STONES

RHODODENDRON MAXIMUM

Subgenus *Hymenanthes*
Section *Ponticum* subsection *Pontica*

A native of eastern North America where it is common and widespread throughout its range, *R. maximum* was introduced into cultivation in Britain by Peter Collinson in 1736, and described by Linnaeus in 1753. It is heat-resistant as well as hardy, tolerates poor soil and heavy shade, but the rather small trusses of purplish pink flowers do not show to advantage, being for the most part hidden by the young leaves which expand in June and July, at, or slightly before, flowering time. The species is rarely grown in gardens, presumably because of its modest display. It is, nevertheless, an important member of the genus, having played a not inconsiderable part in the development of hardy hybrids. In the wild it prefers moist situations, but will tolerate dry ones, forming dense thickets.

The accompanying plate by Sydenham Edwards (tab. 951) was drawn at Messrs Whitley & Brame's of Old Brompton in spring 1785, in a year when their plants of both *R. maximum* and *R. ponticum* produced an abundance of flowers. A clone called 'Summertime', received an Award of Merit at the RHS when exhibited in 1974 by the Crown Estate Commissioners, Windsor.

Rhododendron maximum is most closely allied to *R. macrophyllum* from western North America which has proportionally broader leaves and is eglandular.

Rhododendron maximum Linn., Sp. Plant. ed. 1, 1: 392 (1753).
R. procerum Salisb., Prodr. 287 (1796), *nom. superfl.*
R. maximum L. var. *purpureum* Pursh & var. *album* Pursh, Fl. Am. sept. 1: 297 (1814).
?*R. latifolium* Hoffmanns. in Verz. Pflanzen Kult. Suppl. 2: 195 (1826).
R. purpureum (Pursh) G. Don, *op. cit.* 3: 843 (1834).
R. ashleyi Coker in J. Elisha Mitchell Sci. Soc. 51: 189, t.53, 54 (1935).

Description. A lax, evergreen shrub or small tree up to 3.5 m in height (12 m in the wild) with somewhat rough greyish bark; young growth tomentose, glandular, soon becoming glabrescent. Leaves oblanceolate-elliptic, 10-16(-30) cm long, 3-5(-7) cm wide, with cuneate base and acute or shortly, bluntly cuspidate apex, glabrous above at maturity, with a thin surface film beneath which persists, at least towards the base and near the midrib. Petioles stout, 2-3(-4) cm long, sparsely tomentose. Inflorescence a more or less elongated truss of 14-25 flowers; rachis 10-30 mm long, lengthening in fruit. Pedicels 2-3 cm long, sparsely glandular. Calyx 3-5 mm long, with rounded lobes, glandular. Corolla white, white flushed with pink, pink or deeper purplish pink with yellowish green flecks on top-most lobe, campanulate with rounded lobes, 2.5-3 cm long. Stamens 10, unequal; filaments flattened and hairy at base. Ovary glandular and hairy. Style slender, glabrous. Stigma small. Capsule 1.7-2 cm long.

Distribution. Eastern North America from Nova Scotia to north Georgia.

Habitat. Upland wood and conifer forests, 300-1700 m.

Rhododendron maximum
SYDENHAM EDWARDS

RHODODENDRON MACROPHYLLUM

Subgenus *Hymenanthes*
Section *Ponticum* subsection *Pontica*

Apart from J.D. Hooker, who wrote the text which accompanies Fitch's plate (tab. 4863) I know of no author who has enthused about this plant. Yet, in common with many other rhododendrons, when well grown it is an attractive species which can be depended upon to produce a fine display. The carmine buds opening to yellow-, brown- or red-speckled, pink or rosy purple flowers with crinkly corolla-lobe margins, show to advantage against the ruffs of crowded dark green leaves. *R. macrophyllum* comes into bloom towards the end of the rhododendron season, producing its flowers in May and June when most other species are past their best. There is also a white wild form. Due to a tendency to become leggy when grown in shade, the species is best suited to an open situation on gravelly soil,where it will flower freely even on very small plants and be more likely to retain a relatively compact habit. As would be expected of a plant native to California, *R. macrophyllum* is resistant to drought and heat as well as being hardy, but regrettably it is not easy to propagate from cuttings, so is best raised from seed.

Better known by its synonym, *R. californicum*, this widespread, hardy American species was introduced from California by Veitch's collector William Lobb in 1850. It is one of the only three species occurring in western North America and its range extends from California northwards to British Columbia, along the western side of the continent.

Five years after its introduction, plants of *R. macrophyllum* were exhibited from Veitch's Chelsea nursery, at a flower show in the Sydenham Crystal Palace in June 1855, receiving a mention in the following week's edition of the *Gardeners' Chronicle*.

Rhododendron macrophyllum [D.Don ex] G. Don, Gen. Syst. 3: 843 (1834).
R. californicum Hook. fil. in Bot. Mag. 81: t. 4863 (1855).

Description. Evergreen, tree-like shrub up to 4(-9) m in height with an open habit and stout branches; young growth green and soon glabrous. Leaves often crowded at the tips of the branches, broadly elliptic to oblong, (6.5-)8.5-12(-23) cm long, 3-5.2(-7.5) cm wide, with cuneate base and acute to minutely apiculate apex, glabrous above and beneath at maturity. Petioles 1-2(-3) cm long, glabrous. Inflorescence a condensed raceme of 10-20 or more flowers; rachis 15-30 mm long. Pedicels 30-60 mm long, glabrous, hardly lengthening in fruit. Calyx 5-lobed, about 1 mm long, glabrous. Corolla white to pink or rosy purple flecked with yellow, brown or red, broadly campanulate, 30-40 mm long; lobes usually distinctly wavy or crisped. Stamens 10, unequal, 1.5-2.5 cm long, included within corolla; filaments pubescent in lower third. Ovary densely white- or rufous-hairy. Style slender, exceeding stamens, glabrous. Stigma small, lobulate. Capsule about 25 mm long.

Distribution. A North American species occurring from British Columbia south to California.

Habitat. Edges of conifer forests in sun or semi-shade, or among shorter ericaceous scrub, sea-level to 1200 m.

Rhododendron macrophyllum
WALTER HOOD FITCH

RHODODENDRON ADENOPODUM

Subgenus *Hymenanthes*
Section *Ponticum* subsection *Argyrophylla*

'Exciting' is not a word to be applied to *R. adenopodum* (in Bean (1976) it is dubbed 'dull') but not all growers find it unattractive. Preference is, after all, a personal and subjective matter and the species' lack of universal appeal did not prevent it from receiving an Award of Merit from the RHS in 1926. The plant which won the award was exhibited by Gerald Loder, Wakehurst Place, Sussex. *Rhododendron adenopodum* does have merits. Appearing in April or May, the pink flowers, with or without crimson spotting, are unremarkable, but they are enhanced by the distinctive narrow leaves with grey-felted under surfaces. A slow rate of growth, relatively small stature for the subsection, and spreading habit (it is often wider than high) make it a particularly suitable candidate for the smaller garden.

The French missionary, Farges, discovered *R. adenopodum* in north-east Sichuan and Franchet described it in 1895, using a specific epithet which refers to the glandular pedicels. E.H. ('Chinese') Wilson first introduced the species into cultivation, while collecting for Messrs Veitch in west Hubei in 1900. The following year Farges himself collected seed, from which the French nurseryman Vilmorin raised seedlings and the first of these flowered at Les Barres in 1909. Vilmorin's plants were grown in a glasshouse, but the species is now known to be reliably hardy.

Due to the elongated inflorescence, the species was included for a time in the *Caucasicum* subseries of series *Ponticum*. Subsection *Argyrophylla*, in which it is now placed, includes all the species previously placed in series *Arboreum* subseries *Argyrophyllum*, and Chamberlain (1982) relates this subsection to subsections *Pontica* and *Taliensia*, both morphologically and geographically.

The species has never been figured in *Curtis's Botanical Magazine* and the accompanying plate was painted by Valerie Price especially for this book, from a plant growing at Wakehurst Place, Sussex.

Rhododendron adenopodum Franchet in J. Bot. (Morot) 9: 391 (1895).
R. youngae Fang in Contr. Biol. Lab. Chin. Assoc. Advancem. Sci., Bot. ser. 12: 24 (1939).
R. simiarum Hance subsp. *youngae* (Fang) Chamberlain in Notes Roy. Bot. Gard. Edinburgh 37: 329 (1979).

Description. A spreading evergreen shrub up to 3 m in height, often less, and wider than high; young growth densely tomentose and sparsely glandular. Leaves coriaceous, rather narrow, oblanceolate, 9-18(-20) cm long, 2-4(-5) cm wide, with cuneate base and acuminate to shortly cuspidate apex, at maturity dark green and glabrous above, densely grey- to fawn-felted beneath. Petioles 2-3 cm long, tomentose. Inflorescence a lax, elongated, 6-8-flowered raceme; rachis 1-1.5 cm long. Pedicels 3-4 cm long, stipitate-glandular. Calyx 3-6(-15) mm long, unequally 5-lobed, the lobes oblong, membranous, glandular-ciliate. Corolla pale rose-pink, usually with crimson-purple spots, funnel-campanulate, 4.2-5 cm long. Stamens 10, unequal, as long as corolla; filaments pubescent towards base. Ovary densely fulvous stipitate-glandular. Style slender, glabrous. Capsule cylindrical, about 1.5 cm long, in persistent calyx.

Distribution. China (east Sichuan, Hubei).

Habitat. In thin woods among rocks, 1500-2200 m.

Rhododendron adenopodum
VALERIE PRICE

RHODODENDRON THAYERIANUM

Subgenus *Hymenanthes*
Section *Ponticum* subsection *Argyrophylla*

The specific name of this *Rhododendron* was chosen to honour the Thayer family of Lancaster, Massachussetts, whose long-standing interest in horticulture and support of the botanical exploration of China has been so influential.

The species is native to the province of Sichuan in China, where it is found in the woodlands of the Moupin region at an altitude of around 2700 to 3000 m. Wilson first discovered the plant in fruit in October 1910, while on his second expedition for the Arnold Arboretum and it was not until a plant raised from that seed (no 4273) flowered, that the full charm of the species was realized. It is not only the freely produced, delicately tinted pink and white, crimson-spotted flowers and deep pink buds which make the plant attractive, but also the narrow leaves bronzed underneath and held stiffly erect above the semi-persistent tawny perulae. This species is hardy, easy to grow and the flowers are produced late in the season (June and July) followed by plentiful good seed, yet despite these obvious merits, *R. thayerianum* is not often grown in gardens. It is commercially available, but for some reason has never become popular.

The plate in *Curtis's Botanical Magazine*, tab. 8983, was painted by Lilian Snelling from material grown by Mr J.C. Williams of Caerhays Castle, Cornwall and was sent to Kew in 1923, having flowered at Caerhays in July of the previous year. Persistent perulae and a glandular style distinguish *R. thayerianum* from its close relative *R. hunnewellianum*.

Rhododendron thayerianum Rehder & Wilson in Sargent, Pl. Wilsonianae 1: 529 (1913).

Description. An evergreen shrub up to 3 or 4 m in height with stout, more or less floccose and glandular branches and persistent perulae, at least on young shoots. Leaves coriaceous, crowded towards the end of the current year's growth, narrowly oblanceolate, 8-13(-15) cm long, 1.5-3(-3.8) cm wide, with long tapering base and cuspidate apex, margins recurved, upper surface glabrous when mature, lower surface with a dense, fawn, thin or compacted, felt-like indumentum. Petioles 1-2 cm long, white-floccose, soon glabrous. Inflorescence terminal, a loose truss of 10-20 flowers; rachis 3-4 cm long, floccose and glandular. Pedicels slender, 2.5-5 cm long, glandular. Calyx 2-5 mm long, unequally divided into 5 oblong, rounded lobes, glandular. Corolla white, white flushed with pink, or pale pink, opening from usually dark pink buds, the upper lobes usually spotted with crimson within, sometimes the lobes with a dark pink median line externally, funnel-shaped or funnel-campanulate, 2.5-3 cm long. Stamens (8-)10, unequal, 1.3-2 cm long; filaments pubescent in lower half. Ovary rufous stipitate-glandular or glandular and rufous-hairy. Style somewhat exserted, glandular throughout its length. Capsule cylindrical, approximately 2 cm long, encrusted with shrivelled glands and enclosed in persistent calyx. Seeds 3-4 mm long, with a rounded crest at each end.

Distribution. China (Sichuan).

Habitat. Woodland, 2700-3000 m.

Rhododendron thayerianum
LILIAN SNELLING

RHODODENDRON RIRIEI

Subgenus *Hymenanthes*
Section *Ponticum* subsection *Argyrophylla*

Typical in most characters of subsection *Argyrophylla* in which it is placed, *R. ririei* deviates from other members of the subsection in producing corollas with nectar-pouches. In this respect therefore it must be considered intermediate between subsections *Argyrophylla* and *Arborea*. As with *R. niveum*, which bears paler mauve inflorescences, not everyone finds the flower colour of *R. ririei* attractive, and the species is rarely seen in gardens. The deep purplish mauve or red corollas with crimson or purple-black nectar-pouches, appear in early spring, when the sunlight is pallid or absent and consequently strike a rather sombre note. But if the flower colour fails to please, the species has other merits. It is hardy and easy to grow, although eventually becomes too large for many gardens; the lower leaf surface is clothed with a plastered, silvery white to grey-green indumentum, and the freely produced flowers are large and long-lasting if protected from frost. Moreover the flowering period is long. Like it or not, the species provides a focal point in the garden in February and March, when it scores as one of the few plants in full bloom.

Rhododendron ririei was discovered by E.H. Wilson in 1904 in west Sichuan, on Emei Shan (Mt Omei), and was introduced by him that same year, through Messrs Veitch. The specific epithet refers to the Revd B. Ririe, who was a friend of Wilson. Subsequently the species was collected by others, but always in the same area, to which it appears to be confined.

The species exhibits well, and when shown at the RHS by Lady Aberconway and the Hon. H.D. McLaren in 1931, it received an Award of Merit.

The accompanying plate by Lilian Snelling was passed for publication in *Curtis's Botanical Magazine*, but was withdrawn in favour of *R. kotschyi* and apparently never considered for publication subsequently. It was painted in 1925 from material supplied by Mr Magor of Lamellen, Cornwall and the following notes concerning this plant were communicated by his son, Mr Walter Magor, in a letter to D.R. Hunt:

> 'I only know of one plant of *R. ririei* in this garden, and this is recorded in my Father's planting book as "1808W Got from James Veitch Nov. 1909". It is planted across the valley from the house, and usually comes out in February, the first species to flower every year, and was about 30 feet high when a big Monterey Pine fell across it about three years ago. It is multi-stemmed however and has survived. My Father crossed it with *R. sutchuenense* and named the resulting hybrid 'Maya' which received an A.M. in 1940, the year before he died.'

Rhododendron ririei Hemsley & Wilson in Kew Bulletin 1910: 111 (1910).

Description. A small, evergreen tree up to 16 m in height with rough bark; young growth scurfy; leaf-buds long, pointed. Leaves coriaceous, elliptic to oblanceolate, 9.5-17 cm long, 3.2-5.2 cm wide, with cuneate to rounded base and acute to shortly acuminate apex, upper surface glabrous, lower surface with thin, plastered, white-grey eglandular indumentum. Petioles stout, 1.5-2 cm long. Inflorescence a lax truss of (4-)7-10 flowers; rachis 3-5 mm long. Pedicels stout, 5-10 mm long. Calyx pink, 1-2(-5) mm long, unequally lobed, with a thin white indumentum. Corolla 4-5 cm long, deep purplish mauve or purplish red or magenta to deep violet, sometimes speckled, nectar-pouches darker in colour and glossy, campanulate with broad-based tube and 5-7 lobes. Stamens usually coloured as corolla, usually 10, unequal; filaments purple, glabrous. Ovary densely grey-felted. Style reddish, glabrous. Stigma capitate. Capsule about 2.5 cm long.

Distribution. China (apparently confined to the Emei Shan (Mt Omei) region of Sichuan, although possibly also occurring in Guizhou).

Habitat. Open, rocky slopes, forests and thickets, 1200-2200 m.

Rhododendron ririei
LILIAN SNELLING

RHODODENDRON ARBOREUM

Subgenus *Hymenanthes*
Section *Ponticum* subsection *Arborea*

Rhododendron arboreum has been accorded good coverage over the years by the contributors to *Curtis's Botanical Magazine*, and a total of eight plates, illustrating forms, varieties and species now considered to be conspecific with *R. arboreum*, have been been published to date. The illustration reproduced here in Plate 86A was prepared by Fitch from a field sketch by Joseph Hooker (Plate 86B) and admirably illustrates Fitch's skill at bringing a plant to life by developing a rough drawing; this particular painting was never published in *Curtis's Botanical Magazine*.

Although it was first reported from Kashmir in 1796 by Captain Hardwicke, *R. arboreum* was not named by Sir James Smith until 1804, ten years or so before it was introduced into cultivation and about twenty years before it first bloomed in cultivation in Europe. This 'first flowering' took place at 'The Grange', Alresford, Hants, the home of Mrs Alexander Baring. Who first introduced the seed, and when, remains in doubt. Possibly Francis Buchanan-Hamilton should be credited with it, but this cannot be ascertained. It is known that Hamilton (previously Buchanan) lived in Nepal in 1802 and 1803, and that he sent seeds of *R. arboreum* from there to Roxburgh at Calcutta Botanic Garden, also that Wallich collected seed of white *R. arboreum* in 1821 when he visited Nepal. It is strange that the introduction of the species is not mentioned by Aiton in *Hortus Kewensis*, the last part of which was published in 1814, yet Hardwicke had sent seeds to England in 1796 or 1797. Presumably this seed failed, as nowhere is there any mention of plants having been raised from it.

When it first appeared on the garden scene, everyone wanted to grow the beautiful 'tree rhododendron', but it was soon discovered that this woodland species was tender, some forms more so than others, and would not survive even normal British winters unprotected. Only in the mildest localities, such as south-west England, could it succeed, and even there, some protection from the weather was necessary. Joseph Hooker is said to have claimed that *R. arboreum* did better in Cornwall than in its native Sikkim. Certainly where the species grows well, it excels in size and number of flowers, and one of the most impressive examples of a successful planting is at Lochinch, Wigtownshire, the home of the Earl of Stair. The tenth Earl planted his garden extensively with *R. arboreum* on the advice of Hooker, who also supplied the seed. By 1917, the garden was 70 acres in extent with large numbers of the resulting plants thriving. In New Zealand it does well enough to be grown as a street tree; both the species and its hybrids are used in situations where the bark supplies an additional attractive feature. As well as the numerous forms of *R. arboreum* in cultivation, there is also a multitude of hybrids which owe their existence to *R. arboreum* 'blood'. One of the most important aims of the early hybridists was to breed hardiness into the *arboreum* group, for this was the one essential quality lacking in an otherwise ideal garden plant. This they achieved, with many modern hybrids offering a range of hardiness to suit most situations. One hybrid in particular should be mentioned — *R. nobleanum*, a cross between *caucasicum* and *arboreum*, (also available as × *pulcherrimum* Lindl.). It is the best hybrid for exceptionally early, if not spectacular, flowers. In a mild winter it will start blooming in November or December, and continue to produce its pink or red blossoms spasmodically through January and into February, even until April in some instances. The cross was made by several growers, including W. Norbiton in 1829 and A. Waterer Snr at Knap Hill in 1932; *R. nobleanum* has the endearing quality of starting to flower at the early age of between three and five years.

Although the hardiness of *R. arboreum*'s many forms is so variable, being derived from a woodland species, they all appreciate a certain amount of protection in the form of shelter from wind and frost and shade from the sun. On the other hand, too much shade should be guarded against, as it causes the plants to become leggy. Pruning is not usually recommended for rhododendrons, but even severe pruning if deemed necessary, will be tolerated by *R. arboreum*, as the species has the power of regeneration, even if cut to the ground!

Previously included in series *Arboreum* subseries *Arboreum*, *R. arboreum* has been placed by Chamberlain (1982) in section *Ponticum* subsection *Arborea*, which includes series *Arboreum* in its entirety and is probably related to subsection *Argyrophylla*. Chamberlain also notes that subsp. *arboreum* apparently merges with subsp. *cinnamomeum* var. *roseum* and with subsp. *delavayi*.

Rhododendron arboreum Sm., Exotic Botany 1: 9, t. 6 (1805) subsp. **arboreum**.
R. puniceum Roxb., Fl. Brit. India 2: 409 (1832).

Description. Tree (1-)5-50 m, usually with single trunk up to about 9-20 cm in diameter (1 m from the ground) in Manipur, according to F. Kingdon Ward, sometimes with multiple trunks. Leaves elliptic, sometimes ovate, (8-)10-19 cm long, (2.4-)3-5 cm wide, upper surface smooth to bullate, usually reticulate, lower surface with plastered or dendroid indumentum, usually white or silver in colour. Petioles 1-2 cm long, glandular with loose, persistent indumentum or becoming glabrescent. Inflorescence dense, 10-20-flowered; rachis 1.5-2 cm long. Pedicels up to 1 cm long, hairy and glandular. Calyx very small, almost obsolete, 1-2 mm long with rounded lobes, sparsely glandular or glabrous. Corolla fleshy, usually bright red or crimson, occasionally pink or white, the upper lobes rarely (in subsp. *arboreum*) speckled with black, tubular-campanulate, 3-5 cm long, with nectar-pouches at the base. Ovary white-tomentose, sometimes glandular also. Capsule slender, 1.5-3 cm long.

Distribution. The geographical distribution of the infraspecific ranks of *R. arboreum* is notoriously complex and has been much discussed over the years in many easily accessible publications, so a brief summary is all that is required here. The species in its widest sense has its main concentration in the Himalaya, with a certain amount of overlap of subspecies and varieties occurring in Nepal and Bhutan. Of them all, var. *delavayi* is the easternmost, extending into China, Burma and Thailand, while the southern limit is indicated by subsp. *nilagiricum* in southern India and subsp. *zeylanicum* in Sri Lanka. Subsp. *arboreum* occurs in north India (Kashmir to Sikkim) Nepal, Bhutan.

Habitat. Open or mixed woodland. The altitudinal range of subsp. *arboreum* is 1850-3200 m.

A. *Rhododendron arboreum*
WALTER HOOD FITCH

B. *Rhododendron arboreum*
JOSEPH DALTON HOOKER

RHODODENDRON NIVEUM

Subgenus *Hymenanthes*
Section *Ponticum* subsection *Arborea*

Unlike *R. arboreum*, this member of subsection *Arborea* is not favoured by every rhododendron enthusiast. Many people dislike the rather unusual flower colour, dubbing it 'dingy', 'dirty', 'dull' or even 'depressing' and certainly some colour forms are much less attractive than others. It is also true to say that the smoky mauve-purple flowers do not glow with colour, but the neat round heads are freely produced and in most colour forms the impression of dinginess is entirely dispelled when the flowers are sunlit. The near-white indumentum of the lower leaf surface makes a good foil for the flowers, as well as providing a strong contrast to the very dark upper surface. All these factors contribute towards making *R. niveum* a charming and restful subject for eyes which have been dazzled by the strident colours of the more spectacular species. *Rhododendron niveum* blooms in April and May. It likes a moist climate and sheltered position, but is hardy in the majority of British gardens. It looks particularly attractive when surrounded by grass.

The species was discovered by J.D. Hooker in 1849, in the Sikkim Himalaya where it is now thought to be dying out, although at the time of Hooker's visit it was not uncommon around Lachen, Lachung and Chola, between 3000 and 3700 m. Hooker introduced it in the year of its discovery and a young plant flowered at Kew in May 1853. The plate reproduced here is tab. 6827 painted by Matilda Smith.

A form of *R. niveum* with Imperial Purple (HCC 33/3-33/2) flowers was exhibited by Mrs R.M. Stevenson of Tower Court, Ascot, at the RHS in 1951, when it received an Award of Merit.

The species has no close allies.

Rhododendron niveum Hook. fil. in Rhodo. Sikkim Himal., conspectus, tab. 4 (1851).

Description. An evergreen tree up to 6 m in height with pale brown or greyish bark; young growth densely white-tomentose. Leaves coriaceous, oblanceolate-elliptic, 11.5-17 cm long, 4-4.5(-6.3) cm wide, with obtuse or tapered base, obtuse or rounded apex, glabrous upper surface and densely white to grey- or fawn-tomentose lower surface. Petioles 1-1.5 cm long, floccose. Inflorescence a neat, round, dense truss of 15-20(-30) flowers; rachis about 2 cm long. Pedicels about 1 cm long, densely white- to reddish-tomentose. Calyx obsolete, 1-2 mm long, obscurely lobed. Corolla ashy mauve, grey-blue, smoky purple or deep magenta-purple with darker nectar-pouches, tubular-campanulate, 3-3.5 cm long with 5 spreading lobes. Stamens 10, unequal, 1.8-2.8 cm long; filaments glabrous. Ovary densely white to fawn-tomentose with a lobulate, glabrous basal disk. Style glabrous. Stigma small. Capsule oblong-cylindrical, about 2 cm long.

Distribution. India (Sikkim), Bhutan.

Habitat. Rocky valleys, mixed forest, 2900-3650 m.

Rhododendron niveum
MATILDA SMITH

RHODODENDRON WASONII

Subgenus *Hymenanthes*
Section *Ponticum* subsection *Taliensia*

Some forms of this plant are yellow-flowered, unusual in the large and complex subsection *Taliensia*, the members of which are predominantly white- or pink-flowered. Previously split into four subseries, Tagg's series *Taliense* and series *Lacteum* have been incorporated by Chamberlain into his subsection *Taliensia*. His reasons for maintaining without division so large and unwieldy a collection of species are given in his revision of subgenus *Hymenanthes* (1982).

Rhododendron wasonii is an attractive species, hardy, easily grown and free-flowering at a comparatively early age. The yellow flowers are flushed and spotted with pink or red, and the smallish leaves are backed with a soft felt, at first whitish, later becoming rusty red and finally dark brown. There is another form of the species in cultivation with whitish or pink flowers (forma *rhododactylum*) which received an Award of Merit from the RHS in 1974 when shown by the Crown Estate Commissioners, Windsor. The status of this form is uncertain and requires further investigation. *Rhododendron wasonii* is confined to central Sichuan in China, but is locally common in the forests of Tatsienlu, growing on cliffs, where it was discovered by E.H. Wilson in 1904 at altitudes of between 2300 and 3000 m. He was also responsible for its introduction.

The Marquis of Headfort supplied the material from which tab. 9190 was prepared by Lilian Snelling. The specimen was taken from a plant of Veitch's stock, itself raised from Wilson's collection. The species is named after Lieutenant-Commander (later Rear-Admiral) Cathcart R. Wason R.N. (1874-1941) of HMS Woodlark, who was helpful to Wilson while based at Chungking in 1903-04; it was formally described in 1910.

Rhododendron wasonii Hemsley & Wilson in Kew Bull. 1910: 105 (1910).

Description. A sturdy, evergreen, sprawling shrub up to 1.5 m in height, occasionally more, with stout, rigid young branches. Leaves coriaceous, ovate-lanceolate, (5-)7-8(-10) cm long, 2.3-4(-4.5) cm wide, with cuneate base and apiculate to shortly acuminate apex, dark glossy green above, clad with a sparse to dense whitish felt becoming red-brown with age, and with a few glands intermixed. Petioles stout, 5-10(-15) mm long, tomentose and sparsely glandular. Inflorescence terminal, a loose to more or less compact, 8-15-flowered corymbose truss; rachis 5-10 mm long. Pedicels (15-)20-30 mm long, tomentose. Calyx usually oblique, about 5 mm long, 5-toothed, tomentose. Corolla yellow, occasionally creamy white, with pink, red or crimson flush and flecks, widely campanulate, 35-40 mm long, minutely pubescent at base within. Stamens 10, unequal, 1-1.7(-2) cm long, included; filaments pubescent in lower half. Ovary densely reddish-hairy, eglandular. Style exserted, glabrous. Stigma capitate, lobulate. Capsule cylindrical, up to 1.9 cm long, clad with detersile brown felt.

Distribution. China (central Sichuan).

Habitat. Cliffs or rocks in forest, 2300-3000 m.

Rhododendron wasonii
LILIAN SNELLING

RHODODENDRON LACTEUM

Subgenus *Hymenanthes*
Section *Ponticum* subsection *Taliensia*

Franchet described this species in 1886 from material collected in 1884 by the discoverer of the plant, Abbé Delavay. Franchet gave it the name '*lacteum*' in reference to the flower-colour as described on Delavay's original label — i.e., 'blanc de lait' (milk white). Later, Delavay gave the flower colour as 'jaune soufre' (sulphur yellow) but by then Franchet's name had been validly published for the (usually) yellow-flowered plant. It is now known that plants of the species with either white or yellow flowers exist and that Stapf's comment that the name '*lacteum*' is misleading was unmerited.

Delavay first found *R. lacteum* near Hoking on Mt Kova La Po in north-west Yunnan, where it was growing at 3200 m in almost pure stands. It was not introduced into cultivation until George Forrest brought it back in 1910 (F 6778) from the Tali range to the south of Delavay's locality and from several other regions subsequently. This hardy, free-flowering rhododendron is a splendid sight when growing well and in full bloom in April and May, but is notoriously difficult to please. It rarely produces seed, propagation by layering is difficult, as is the establishment of a transplanted specimen and the plants have an unfortunate tendency to gradually die back, regardless of any remedial measures taken. Perfect drainage, protection from strong winds and sun, and grafting onto a strong-growing species are known to be beneficial, but do not necessarily guarantee success. Cox (1979) pointed out that many of the healthiest and oldest plants in Britain are flourishing in some of the coldest gardens, but the significance of this fact has apparently not been investigated.

A clone of *R. lacteum*, with the flower colour described as 'sulphur white, with a dark crimson blotch', was awarded a First Class Certificate when exhibited in 1926 by A.M. Williams, Warrington Park, Cornwall. Another First Class Certificate was awarded by the RHS in 1965 to the clone 'Blackhills' which had unmarked primrose-yellow flowers, exhibited by S.F. Christie of Blackhills, Elgin, Morayshire.

Rhododendron lacteum, previously a member of Tagg's series *Lacteum* and Sleumer's subsection *Lactea* has been transferred by Chamberlain (1982) for complex reasons to subsection *Taliensia*. In short, the indumentum characters which separated the two groups were considered to be unsatisfactory.

Tab.8988 was painted in 1922 by Lilian Snelling from material supplied by Mr Williams, Caerhays, Cornwall.

Rhododendron lacteum Franchet in Bull. Soc. Bot. France 33: 231 (1890).
R. mairei Lév. in Feddes Repert. 12: 285 (1913).

Description. An evergreen shrub or small tree up to 7.5(-9) m in height with purplish brown bark and stout branches. Leaves coriaceous, elliptic-obovate, 8-17(-18.5) cm long, 4.5-7(-10) cm wide, with rounded to cordulate base and rounded, apiculate apex; upper surface glabrous at maturity, lower surface clothed with a thin, unistrate, fawn or grey-brown, suede-like indumentum of radiate hairs. Petioles 2-2.5(-4.7) cm long, glabrescent. Inflorescence a dense, terminal truss of 15-30 flowers; rachis 2.5-3 cm long. Pedicels 2.5-3 cm long, at first floccose, soon glabrescent. Calyx an undulate or shallowly lobed glabrous rim about 1 mm long. Corolla clear yellow, rarely white, unmarked or with a basal purple blotch, widely campanulate, 4-5 cm long. Stamens 10, unequal, included; filaments puberulous towards base. Ovary densely tomentose. Style glabrous. Stigma small, lobulate. Capsule curved, about 2 cm long.

Distribution. Apparently confined to China (west Yunnan), where it is locally common.

Habitat. Forest margins and rocky slopes, (3200-)3700-4000 m.

Rhododendron lacteum
LILIAN SNELLING

RHODODENDRON FULVUM

Subgenus *Hymenanthes*
Section *Ponticum* subsection *Fulva*

Subsection *Fulva* is comprised of two species, *R. uvarifolium* and *R. fulvum*, which have in common an unmistakable narrow, curved ovary which develops into a sickle-shaped capsule, and a granular indumentum unlike that of any other rhododendron. *Rhododendron fulvum* now includes *R. fulvoides*, which is really only a narrower-leaved variant with a dark or yellowish brown indumentum. The subsection has no apparent allies, but may be related to subsection *Taliensia* or *Argyrophylla*. George Forrest was the first person to collect *R. fulvum*, although others re-introduced the species several times. Forrest found it in August 1912, at an altitude of 3000 to 3300 m on the western flank of the Shweli-Salween divide where he gathered seed from which plants were raised. Later he, too, re-introduced the species many times and it has proved to be completely hardy in the British Isles. The subject of Lilian Snelling's painting (tab. 9587) was supplied by the Marquess of Headfort, Kells, Co. Meath in March 1938. The plants he raised produced three colour forms, but such variation is not unusual in naturally widespread species, indeed, the colour of the leaf indumentum is similarly variable and can be pale fawn, bright cinnamon or rich red-brown. The species is not one which is likely to excite the grower. Nevertheless, *R. fulvum* is an attractive, dependably hardy creature for gardens large enough to accommodate its potential height of 9 m or so, in woodland. It becomes free-flowering when quite young, producing its blooms in March and April, and as has been indicated, is a good foliage plant. In cold, frosty weather, the leaves droop and the margins curl inwards until the under surface is completely hidden, so the whole plant looks decidedly miserable, but this is merely a temporary defence mechanism and when the cold recedes, the plant usually returns to normal quite rapidly.

The RHS showed appreciation of the species by giving an Award of Merit to a pink-flowered form exhibited by the Hon. H.D. McLaren of Bodnant in 1933.

Rhododendron fulvum Balf. fil. & W.W. Sm. in Notes Roy. Bot. Gard. Edinburgh 10: 110 (1917).
R. fulvoides Balf. fil. & Forr. in Notes Roy. Bot. Gard. Edinburgh 12: 112 (1920).

Description. An evergreen shrub or small rounded tree, 2-8 m (to 9 m in cultivation) in height, with rough bark; young growth rufous-tomentose. Leaves coriaceous, oblanceolate-elliptic, 8-22(-25) cm long, 3.6-9 cm wide, with cuneate base and rounded, apiculate apex, upper surface glabrous, lower surface with a dense bistrate, suede-like indumentum composed of mop-like hairs which have a granular appearance. Petioles 1-2(-4) cm long, glandular and tomentose. Inflorescence a dense compact truss of 10-20 flowers; rachis (6-)10-15(-22) mm long. Pedicels slender, (1-)2-3 cm long, glabrous. Calyx obsolete, about 0.5(-2) mm long, glabrous. Corolla white to pink, usually with a crimson basal blotch, sometimes also flecked with crimson, campanulate, 2.5-4.5 cm long. Stamens 10, unequal, 1.7-2.7(-3.5) cm long. Ovary curved, narrow, with lobulate basal disk, glabrous. Style slender, glabrous. Stigma small, discoid, lobulate. Capsule sickle-shaped, 2.5-4 cm long.

Distribution. China (west Yunnan, south-east Xizang) and adjacent north-east Upper Burma.

Habitat. Forests, cliffs and rocky slopes, 2400-4700 m.

Rhododendron fulvum
LILIAN SNELLING

RHODODENDRON LANATUM

Subgenus *Hymenanthes*
Section *Ponticum* subsection *Lanata*

Rather surprisingly, this handsome species has never been figured for *Curtis's Botanical Magazine*, and the illustration reproduced here was painted by Walter Fitch for another prestigious publication, J.D. Hooker's *Rhododendrons of the Sikkim Himalaya*. No doubt the omission was at least partly due to the temperamental behaviour of the species in cultivation. It is hardy, but extremely difficult to grow successfully; it is slow-growing, shy-flowering and readily succumbs to the first signs of drought. A sheltered position with a plentiful supply of leaf-mould is essential to its well-being. Frequent mulching is another requirement if this perverse species is to flourish. If all efforts to please it, plus a little perseverance, produce a well-grown plant of *R. lanatum*, the grower is amply rewarded, for it is a lovely species. The large flowers are cream to yellow, sometimes fading to white lightly spotted with red or crimson and the under surface of the leaves is covered with a dense, woolly, usually coffee-brown indumentum, so that even when not in flower, the plant still supplies a focal point in the garden and is well worth growing for the foliage alone. But when the plant is in full bloom in April and May, the contrasts between the flowers, the very dark green upper leaf surfaces and the coffee-brown under surfaces provide a breathtaking spectacle. The indumentum is dense and thick enough to be peeled off, twisted and used as lampwicks in its native habitat by the local people.

Joseph Hooker discovered and collected the species in 1848 in Sikkim and introduced it into cultivation in 1851. Within the altitudinal range of 3000 to 4500 m the species is usually found on steep slopes, where it attains 3 m in height, but in cultivation it rarely exceeds 2 m. After Hooker's discovery, the range of *R. lanatum* was extended by other collectors who found it also in Bhutan, Arunachal Pradesh and south Xizang. Chamberlain (1982) commented on the geographical variability of the species and noted that, where the two species grow together, *R. lanatum* apparently hybridizes with *R. tsariense*. As regards relationships, he suggested that perhaps subsection *Lanata* is closer to subsection *Taliensia* than to subsection *Campanulata* (it was previously placed in series *Campanulatum* sensu Tagg).

At the same time, Chamberlain reduced the closely allied *R. luciferum* (Cowan) Cowan and *R. flickii* Davidian to synonyms of *R. lanatum*, though this has been recently questioned. Further research into the classification of *R. lanatum* and its allies is evidently required.

Rhododendron lanatum Hook. fil., Rhodo. Sikkim Himal., 17, t. 16 (1849).

Description. An erect, evergreen shrub up to 4(-7.5) m in height with greyish purple bark; young growth densely rufous-woolly. Leaves coriaceous, elliptic-obovate, 6.5-12 cm long, 2.5-4.2(-5) cm wide, with rounded base and rounded, apiculate apex, upper surface finally glabrous or with persistently floccose midrib, lower surface with a dense, whitish tomentum which becomes coffee-brown or reddish brown. Petioles about 1(-2.6) cm long, densely tomentose. Inflorescence a loose, terminal truss of 5-10 flowers; rachis 3-10 mm long. Pedicels l-2 cm long, densely whitish- to brown-tomentose. Calyx about 1 mm long. Corolla creamy, or pale sulphur-yellow to daffodil-yellow, (sometimes fading to white as it ages) lightly spotted with red or crimson, campanulate, 3.2-5 cm long. Stamens 10, unequal; filaments pubescent towards base. Ovary densely rufous-tomentose. Style glabrous. Stigma lobulate. Capsule curved, 1.5-2.5 cm long.

Distribution. North India (Arunachal Pradesh, Sikkim), Bhutan and China (south Xizang).

Habitat. Forests and scrub in gullies in rocky mountain spurs, on cliffs and rocks, but almost always on steep slopes, 3000-4500 m.

Rhododendron lanatum
WALTER HOOD FITCH

RHODODENDRON WALLICHII

Subgenus *Hymenanthes*
Section *Ponticum* subsection *Campanulata*

Originally considered worthy of specific status by J.D. Hooker, but for a time regarded by him as a variety of *R. campanulatum*, *R. wallichii* has been retained at specific rank by most authors, including Tagg, Millais, and more recently Chamberlain, in his 1982 revision of subgenus *Hymenanthes*. In that revision, the latter author suggested that subsection *Campanulata* is possibly allied to subsections *Lanata* and *Taliensia*. *Rhododendron wallichii* is distinguished from *R. campanulatum* by the differences in the indumentum; the hairs on the lower leaf-surface are dense and capitellate to ramiform in *R. campanulatum*, but sparse and fasciculate in *R. wallichii*. Chamberlain noted that the latter species apparently replaces the former in the 'more humid areas at slightly lower altitudes on the outer slopes of the C. Himalayas'. The species is named after Nathaniel Wallich (1786-1854), sometime Superintendent of the Calcutta Botanic Garden. Joseph Hooker discovered *R. wallichii* in the Sikkim Himalaya and described it in 1849, the year that he also introduced it into cultivation. Subsequently it was found in other nearby countries and re-introduced by several collectors. It was usually growing in forest, but was also often found in open situations.

Fitch's drawing of this hardy, easily grown species (tab. 4928) does justice to a lovely rhododendron. The loose trusses of large lilac or lavender flowers against the very dark green of the mature foliage, together with the crimson perulae and petioles make a particularly pleasing display in a woodland setting during March, April and May. A white form has been found in the wild, but is not as attractive as the more frequent mauve-flowered type.

Tab. 4928 was painted from plants which flowered at Kew in May 1856, and had been raised from Himalayan seed; another plate, tab. 3759, features *R. campanulatum* and was prepared before *R. wallichii* was confirmed at specific level. Tab. 4928 appears as *R. campanulatum* var. *wallichii* with the following note: 'When Dr Hooker prepared his description on its native mountain he considered this to be a new species, which he named after our lamented friend Dr Wallich, not being then aware of the sportive nature of *Rhododendron campanulatum*, of which he afterwards, and no doubt justly, considered it to be a mere variety....'. Chamberlain makes no mention of the variety. In the garden, many growers prefer *R. campanulatum*, which they consider to be superior to *R. wallichii*, due to its continuous, persistent leaf-indumentum, an extra feature of interest when the plant is not in flower.

Rhododendron wallichii Hook., Rhodo. Sikkim Himal., tab.5 (1849).

Description. An evergreen shrub up to 4.5(-6) m in height; young shoots greyish-tomentose, glabrous when mature. Perulae red to crimson. Leaves coriaceous, elliptic, oblong or ovate, 7-11(-14) cm long, (2.5-)3.5-5.8(-6.5) cm wide, with rounded to cordate base and rounded, apiculate apex; dark green, glabrous and often glossy above, lower surface with a sparse, patchy indumentum of dark brown, fasciculate hairs. Petioles stout, often stained red, (0.4-)1-2(-2.8) cm long, floccose. Inflorescence a loose 5-8(-10)-flowered raceme; rachis 1-2 cm long. Pedicels 1-1.5(-2.5) cm long, glabrous or sparsely fasciculate-hairy. Calyx shallowly, unequally 5-lobed, lobes 1-3 mm long, sparsely tomentose or glabrous. Corolla white, white flushed mauve, pinkish or pale mauve, lilac or deeper lavender-purple, with or without speckling on upper lobes, funnel-campanulate, 2.5-4(-5) cm long. Stamens 10, 2.5-3.5 cm long; filaments puberulous at base. Ovary glabrous or nearly so. Style glabrous. Stigma small, lobulate. Capsule 1.5-3 cm long.

Distribution. East Nepal, Bhutan, north India (Sikkim, Bengal) Bhutan and China (south Xizang).

Habitat. Mixed forest, rhododendron scrub and open hillsides, 2900-4300 m.

Rhododendron wallichii
WALTER HOOD FITCH

RHODODENDRON GRIERSONIANUM

Subgenus *Hymenanthes*
Section *Ponticum* subsection *Griersoniana*

The sole member of subsection *Griersoniana*, this species appears to have no close relatives. Traditionally it was included in series *Auriculatum*, but apart from the sticky glands which are common to both groups, there is no real affinity. *Rhododendron griersonianum* is apparently confined to a relatively small area of west Yunnan, where it was first discovered by George Forrest in June 1917. It was growing near the Burma border, in open situations in both pine and mixed forest on the Shweli-Salween divide at an altitude of 2700 m. He collected seed (no 15815), thereby introducing the plant into cultivation in that same year. Subsequently, Forrest re-introduced the species many times, but always from the same small area of Yunnan. In the wild the plants usually grow singly, rather than in groups. *Rhododendron griersonianum* is not only free-flowering, but blooms rather later in the season from June to July (to August if growing in shade) when the majority of species have finished flowering, but plants raised from seed flower at an early age, usually within six years of sowing; the flowers themselves are large, more or less nodding and are borne in a loose truss of five to twelve. In colour they are commonly a bright geranium-scarlet, often with darker speckling, but vary from bright rose through vermilion to soft crimson. If there is a drawback, then it is that the species is from low elevations and is therefore variable in hardiness. It flourishes in Cornwall and other mild parts of Britain, and given the opportunity and a sheltered position, may well succeed in colder areas, particularly as it becomes hardier with age. Hot, dry conditions are tolerated, as are shade and pruning. Altogether an ideal subject for the hybridist, *R. griersonianum* proved to be one of Forrest's most memorable introductions, becoming a remarkably successful parent and giving rise to well over one hundred floriferous hybrids, almost one third of which have received awards from the RHS. The species itself, when exhibited at an RHS show in 1924 by T.H. Lowinsky of Sunninghill and Lionel de Rothschild of Exbury, was awarded a First Class Certificate. The plant from which Lilian Snelling painted tab. 9195, was raised at Kew from seed (Forrest 874) sent by Lionel de Rothschild in 1920.

Rhododendron griersonianum Balf. fil. & Forr. in Notes Roy. Bot. Gard. Edinburgh 11: 69 (1924).

Description. An evergreen, setulose-glandular (sticky) shrub up to 3 m in height, with rough bark; young growth glandular and whitish-tomentose. Perulae long, linear, cuspidate. Leaves elliptic, 10-20 cm long, (2-)3-5 cm wide, with broadly cuneate to rounded base and acute-acuminate apex, dull green, matt and finally glabrous above, densely whitish- to pale brown-tomentose beneath. Petioles 1-2(-3.5) cm long, densely glandular and floccose, at least when young. Inflorescence a loose truss of 5-12 nodding flowers; rachis 1.2-2(-4) cm long. Pedicels 2-3(-4) cm long, densely long stipitate-glandular. Calyx obsolete, 1 mm long. Corolla geranium-scarlet or deep rose to crimson, with or without speckles, tubular to funnel-campanulate, 5.5-8 cm long; tube externally densely hairy and sparsely glandular. Stamens 10, unequal, included; filaments up to 5.7 cm long, pubescent in lower half. Ovary densely white-hairy with scattered glands. Style deep red, exceeding stamens, glabrous except for extreme base, which is glandular and floccose. Stigma discoid, lobulate. Capsule more or less straight, 2(-3) cm long.

Distribution. China (Yunnan, near the frontier with Burma).

Habitat. Open situations in both coniferous and mixed forests, 2100-2700 m.

Rhododendron griersonianum
LILIAN SNELLING

RHODODENDRON KYAWI

Subgenus *Hymenanthes*
Section *Ponticum* subsection *Parishia*

This species was described in 1914 from specimens collected in the south-east Kachin Hills by Maung Kyaw (pronounced Chaw) of the Burmese Forest Service, for whom it is named. At first the species was thought to be confined to north-east Upper Burma, but it is now known to occur in west Yunnan also. It was introduced into cultivation by George Forrest, who collected seed in May 1919 (no 17928). His introduction only marginally preceded that of Farrer & Cox who collected seeds at the type locality near the Hypepat bungalow, a little more than 48 km south of Forrest's locality, in November of the same year.

Being tender, *R. kyawi* flourishes only in milder regions, but when growing well is a splendid species, producing large trusses of bright pink or crimson flowers over a long flowering period lasting from June to August. This means that growth is made rather late in the summer and damage from early frosts before the wood is sufficiently ripened to withstand them, is an ever-present possibility. *Rhododendron kyawi* is not a plant for smaller gardens — it can attain 9 m in height! Although it is frequently recommended as a cool-house subject, the cool-house too, would have to be sizeable to accommodate a fully grown plant of this species.

In the recent Edinburgh revision, subsection *Parishia*, which is probably allied to subsections *Irrorata* and *Neriiflora*, includes *R. agapetum* and *R. prophantum* which Chamberlain has reduced to synonymy under *R. kyawi*. There has been considerable doubt for many years as to the distinctness of these two species and Chamberlain's classification has confirmed a widespread opinion that only one entity is involved, for which the correct and valid name is *R. kyawi*. The painting reproduced here is by Lilian Snelling (tab. 9271), published in *Curtis's Botanical Magazine* in 1929.

Rhododendron kyawi Lace & W.W. Sm. in Notes Roy. Bot. Gard. Edinburgh 8: 216, t. 141 (1914).
R. agapetum Balf. fil. & K. Ward in Notes Roy. Bot. Gard. Edinburgh 13: 58 (1920).
R. prophantum Balf. fil. & Forr. in Notes Roy. Bot. Gard. Edinburgh 13: 58 (1920).

Description. An evergreen shrub or small tree up to 9 m in height with rough bark; young growth stellate-hairy and glandular-setose. Leaves coriaceous, elliptic-oblong, 9-22(-30) cm long, 4-9(-11) cm wide, with rounded base and rounded (sometimes more or less acuminate) apex, upper surface glabrous, lower surface with detersile, reddish brown, floccose indumentum and scattered glands, or more or less glabrescent or becoming so. Petioles (1-)2.5-4(-6) cm long, stellate- and glandular-hairy, finally glabrescent. Inflorescence a 10-16(-20)-flowered raceme; rachis up to 40 mm long. Pedicels 20-30 mm long, glandular-setose and sparsely floccose. Calyx shallowly cupular, 1-2 mm long, glandular-setose with 5 irregular, broadly rounded lobes. Corolla fleshy, bright rose pink, scarlet or bright crimson with darker nectar-pouches, usually unspotted, tubular-campanulate, 46-60 mm long, externally downy. Stamens 10, unequal; filaments often crimson, 3-4 cm long, minutely pubescent in lower half. Ovary densely tawny stellate-hairy and glandular-setose. Style usually crimson, about 4.5 cm long, stipitate-glandular and floccose, at least in the lower half. Stigma discoid, lobulate. Capsule slightly curved, 25-40 mm long.

Distribution. China (west Yunnan), north-east Upper Burma.

Habitat. Thickets, forests, deep wooded gorges and on limestone cliffs, 1500-3700 m.

Rhododendron kyawi
LILIAN SNELLING

RHODODENDRON SMITHII

Subgenus *Hymenanthes*
Section *Ponticum* subsection *Barbata*

In the revision prepared at the Royal Botanic Garden Edinburgh, Chamberlain (1982) retained this species at specific level, relegating *R. argipeplum* to synonymy. Davidian, however, gives both *R. argipeplum* Balf. fil. & Cooper and *R. macrosmithii* Davidian (*R. smithii* Nutt. ex Hook. fil.) specific status. Respective authors cite tab. 5120 as representing *R. smithii* and *R. macrosmithii*. Time and space do not permit either discussion or research of the correct taxonomy or nomenclature here, and the facts are merely noted. However it is clear that the name *R. smithii* applies to a hybrid described by Sweet in 1830. If both *R. argipeplum* and *R. macrosmithii* are maintained then it is probable that tab. 5120 in the *Botanical Magazine* (the illustration presented here) refers to the former, not the latter. Most authors are agreed that *R. smithii* is very closely related to *R. barbatum*, differing mainly in the indumentum of the lower leaf surface, which in *R. smithii* forms a continuous brownish or white layer, (white in *R. argipeplum*) over the whole surface, but in *R. barbatum* is comprised of scattered dendroid hairs and stipitate glands. In cultivation *Rhododendron smithii* is often of a more compact form than *R. barbatum*.

The affinities of the subsection are with subsections *Parishia* and *Thomsonia*. *Rhododendron smithii* was introduced into cultivation by Booth, who discovered the plant on the northern slopes of the Lablung Pass in the Kamang Division of Arunachal Pradesh in north-east India, where it was growing with *R. hookeri*. It flowered first at Nutgrove, Rainhill, Lancashire in 1859, and a drawing was made by Mr Holden of Warrington; tab. 5120 by Fitch, was prepared from Holden's drawing and published in the same year. The plant has been collected several times subsequently in Bhutan and in east Sikkim, where in some areas it is common in *Abies* and *Rhododendron* forests at altitudes between 2400 and 3660(-4000) m. It was named after Sir J.E. Smith, the English botanist who founded the Linnean Society.

When not destroyed by frost, this species which flowers in March and April, provides a lovely display, with trusses of glowing scarlet flowers and copper or plum-coloured young growth, but it is not a plant for the smaller garden, as it often attains 7 m or more in height.

A clone of *R. smithii* named 'Fleurie', received an Award of Merit from the RHS in 1978 when exhibited by R.N.S. Clarke of Borde Hill.

Rhododendron smithii (Nutt. ex) Hook. fil. in Bot. Mag. 85: t. 5120 (1859).
R. argipeplum Balf. fil. & Cooper in Notes Roy. Bot. Gard. Edinburgh 9: 213 (1916).

Description. An evergreen shrub or small tree up to 7.5 m in height, with setose young growth, the setae long and stiff. Leaves coriaceous, convex, elliptic to obovate-lanceolate, 8-15(-20) cm long, 2.7-4(-10) cm wide with rounded base, acute to rounded apex and glabrous upper surface; lower surface with a thin continuous layer of pale brown or whitish dendroid hairs, and often sparsely setose towards the base of the midrib. Petioles stout, 1-2 cm long, densely glandular-bristly. Veins deeply impressed. Inflorescence a tight (10-)15-20-flowered truss; rachis about 5 mm long. Pedicels 1-1.5 cm long, glandular-setose. Calyx fleshy, usually red or stained red, 5-10 mm long, with 5 rounded, subequal and glandular-ciliate lobes. Corolla fleshy, tubular-campanulate, clear scarlet to crimson with darker nectar-pouches, 3-4.5(-5) cm long. Stamens 10, 2.5-3 cm long; filaments glabrous. Ovary densely rufous-hairy and glandular-setose with an undulate to lobed, glabrous basal disk. Style exceeding stamens, glabrous. Stigma small, lobulate. Capsule more or less 1.5 cm long, straight.

Distribution. North India, China (south Xizang), Nepal and Bhutan.

Habitat. Open slopes amongst scrub, and in *Abies* and *Rhododendron* forests, 2400-4000 m.

Rhododendron smithii
WALTER HOOD FITCH

RHODODENDRON SANGUINEUM subsp. SANGUINEUM var. HAEMALEUM

Subgenus *Hymenanthes*
Section *Ponticum* subsection *Neriiflora*

This species is included not so much for the beauty of the plant, which is debatable, but rather for the extraordinary flower colour of the variety, a somewhat glaucous dark crimson, which appears almost black in reflected light.

Originally ranked as a species, *R. haemaleum*, this was reduced to varietal status under *R. sanguineum* by Chamberlain in 1979. Stapf, in his account of *R. sanguineum* for *Curtis's Botanical Magazine* published in 1932, had already reached the conclusion that *R. haemaleum* could not be satisfactorily maintained at specific level and had provisionally reduced it to a synonym of *R. sanguineum*. He analysed the case in some detail, but left the final decision until a formal revision of the whole section *Sanguineum* be undertaken. The species, *R. sanguineum*, now belongs to the taxonomically difficult subsection *Neriiflora* which encompasses Tagg's series *Neriiflorum* and is related to subsection *Thomsonia*. Var. *haemaleum* differs from typical *R. sanguineum* merely in flower colour. It was discovered by George Forrest on the Mekong-Salween watershed at some time between 1904 and 1918, growing with the type. The accompanying plate was painted by Lilian Snelling from plants grown by the Marquis of Headfort, Kells, Co. Meath, which had been raised from Forrest's seed (F 16736) and depicts a particularly dark form, although the colour varies from this to deep claret red.

The mostly low-growing *R. sanguineum* is completely hardy and flourishes in open, rocky or boggy situations, which makes it an ideal subject for the rock-garden or pond-side. A group of mature plants of differing colour-forms, with the sun shining through the flowers, produces a patch of vibrant, glowing colour from March to May. Not all forms are free-flowering, but they do tend to become more so as they mature.

Rhododendron sanguineum Franchet subsp. **sanguineum** var. **haemaleum** (Balf. fil. & Forr.) Chamberlain in Notes Roy. Bot. Gard. Edinburgh 37: 334 (1979).
R. haemaleum Balf. fil. & Forr. in Notes Roy. Bot. Gard. Edinburgh 11: 71 (1917).
R. sanguineum Franchet subsp. *haemaleum* (Balf. fil. & Forr.) Cowan in Notes Roy. Bot. Gard. Edinburgh 20: 69 (1940).
R. sanguineum Franchet subsp. *mesaeum* [Balf. fil. ex] Cowan *ibid.* 20: 70 (1940).

Description. A compact (spindly in shade) dwarf, evergreen or semi-deciduous shrub up to 1.5(-1.8) m in height, with slender branches; young growth sparsely floccose, rarely with eglandular setae also. Perulae persistent or deciduous. Leaves coriaceous, elliptic-obovate, 3-8 cm long, 1.5-3.2 cm wide, with a more or less cuneate base and rounded, apiculate apex; glabrous above, lower surface with a continuous, compacted, silvery grey or fawn, rosulate indumentum. Petioles 5-8 mm long, at first floccose, rarely also glandular, usually becoming glabrous. Inflorescence a loose, terminal cluster of 3-6 flowers; rachis up to 5 mm long. Pedicels slender, 1-2.5 cm long, stipitate-glandular. Calyx red, 3-10 mm long, cupular with 5 rounded, glandular-ciliate lobes. Corolla fleshy, deep blackish crimson, shortly tubular-campanulate with basal nectar-pouches, 2.5-3.8 cm long. Stamens 10, unequal; filaments glabrous or papillate. Ovary tomentose to stipitate-glandular. Style widening towards stigma, glabrous. Stigma small, discoid, lobulate. Capsule 1-1.5 cm long.

Distribution (of the whole species) China (south-east Xizang, north-west Yunnan).

Habitat. Widespread in open, rocky or boggy places or in scrub, 3000-4500 m.

Rhododendron sanguineum subsp. *sanguineum*. var. *haemaleum*
LILIAN SNELLING

RHODODENDRON FORRESTII subsp. FORRESTII

Subgenus *Hymenanthes*
Section *Ponticum* subsection *Neriiflora*

A more familiar name for this dwarf rhododendron is *R. forrestii* var. *repens*, which despite being reduced to synonymy by Cowan and Davidian in 1951, still persists in gardens and the trade.

George Forrest discovered the species in 1905, growing on moss-covered rocks on the Mekong-Salween divide in Yunnan, but thanks to the activities of the rebel lamas, at whose hands he narrowly escaped death, Forrest's collections were all lost, with the exception of one small specimen which had already been despatched home. From this scrap Diels described the species in 1912. Forrest collected the plant again in the same area in 1914 and 1918, but whereas the original specimen had purple undersides to the leaves, the leaves of plants in the two subsequent collections were mostly green beneath. These green-leaved plants were at first described as a new species, *R. repens*, but this was reduced to a synonym of *R. forrestii* by Stapf in 1929. Cowan and Davidian (1951) did not accept this judgement, preferring to maintain *R. repens* as a variety of *R. forrestii*. Chamberlain (1982) split *R. forrestii* into two subspecies as follows:

Lower epidermis of leaf purple or green, not papillate, stipitate glands few or absent; leaves 1.1-1.5(-2.2) times as long as broad · · · · · · · · · · subsp. *forrestii*
Lower epidermis of leaf glaucous-papillate, stipitate glands conspicuous; leaves 2.2-2.6(-3.2) times as long as broad · · · · · · · · · · subsp. *papillatum*

This slow-growing, creeping and carpeting rhododendron with tiny leaves and remarkably large, glowing crimson flowers is a most attractive plant, which when flourishing, creates a small but spectacular display close to the ground, rivalling the flamboyance of the grander species. It is best grown in shady positions in shallow peaty soil on a rock base, on tree-stumps or over a wall so that it can hang down the vertical face, rooting in cracks as it progresses, just as it does in the wild. It flowers in April and May, but not always freely, and the formation of flower-buds can be encouraged by careful siting, thus ensuring that ample light and moisture, as well as protection from spring frosts, are available.

Tab. 9186 was painted by Lilian Snelling from material supplied by the Marquis of Headfort, Kells, in whose garden the plant flowered in March 1926. Under the name *R. repens*, subsp. *forrestii* received a First Class Certificate from the RHS when exhibited in 1935 by J.B. Stevenson of Tower Court, Ascot.

Rhododendron forrestii [Balf. fil. ex] Diels in Notes Roy. Bot. Gard. Edinburgh 5: 211 (1912) subsp. **forrestii**.
R. repens Balf. fil. & Forr. in Notes Roy. Bot. Gard. Edinburgh 11: 115 (1919).
R. forrestii Diels var. *repens* (Balf. fil. & Forr.) Cowan & Davidian in Rhododendron Year Book 6: 69 (1951).

Description. Dwarf, creeping, evergreen shrub up to 30 cm in height, but usually not exceeding 15 cm, the gnarled stems up to 60 cm long, with persistent perulae. Leaves stiff, coriaceous, obovate-orbicular, 1-2.8 cm long, 0.9-1.8 cm wide, base broadly cuneate, often narrowly decurrent, apex rounded to retuse, sometimes mucronate, margin more or less recurved, both surfaces glabrous or lower surface with a few stipitate glands and branched hairs towards the base, green when mature. Petioles 5-8 mm long, stipitate-glandular and sparsely floccose. Inflorescence terminal, 1(-2)-flowered. Pedicels 1-2 cm long, stipitate-glandular. Calyx about 1 mm long, with 5 fleshy lobes. Corolla fleshy, bright crimson, tubular-campanulate, 3-3.8 cm long, with 5 basal nectar-pouches, glabrous. Stamens 10; filaments shorter than corolla, glabrous. Ovary densely stipitate-glandular and reddish-tomentose with a basal, lobulate, puberulous disk. Style glabrous. Stigma as wide as style, lobulate. Capsule 1.5-2 cm long, warty, with persistent calyx.

Distribution. China (north-west Yunnan, south-east Xizang) and adjacent north-east Upper Burma. Subsp. *papillatum* is found in south-east Xizang.

Habitat. On mossy rocks and boulders in moist, stony pasture, 3050-4500 m.

Rhododendron forrestii subsp. *forrestii*
LILIAN SNELLING

RHODODENDRON FLOCCIGERUM

Subgenus *Hymenanthes*
Section *Ponticum* subsection *Neriiflora*

The plant which modelled for Lilian Snelling's painting (tab. 9290) in *Curtis's Botanical Magazine* was grown in the Temperate House at Kew and was a particularly colourful form of a variable species which is widely grown in cultivation. Regrettably the flower colour is rarely as striking as that portrayed here, the plants frequently producing blooms in very dull or muddy shades. The flowers are nodding and not invariably are hidden by the leaves, the undersurfaces of which are covered by a loose, patchy, rufous wool which gives the species its name. In spite of such drawbacks the would-be grower should not be deterred, for this is still a good species for the smaller garden, being hardy, early flowering (March to April) with a relatively dwarf, neat habit and it is not fussy with regard to soil. Bearing in mind its shortcomings, the plant is best bought while in flower, thus making sure that a plant with the desired combination of characters is obtained.

The French missionary, Soulié, discovered *R. floccigerum* near Tseku on the Mekong-Salween divide in April 1895 (no 1014), but it was George Forrest who introduced it into cultivation in 1914, sending seeds home from north-west Yunnan where the species is quite common. He found it in several different locations growing in pine forest, and in open situations on limestone cliffs or among limestone rocks. Subsection *Neriiflora* in all totals 26 species and is allied to subsection *Thomsonia*. *R. floccigerum* is most closely related to *R. neriiflorum* and *R. sperabile*, from which it is distinguished by its discontinuous floccose leaf indumentum.

Rhododendron floccigerum Franchet in J. Bot. (Morot) 12: 259 (1898).

Description. A small, rounded, (leggy in shade) evergreen shrub 0.6-3 m in height; young growth densely floccose, setose-glandular or eglandular. Leaves narrowly elliptic to oblong or elliptic, (3.5-)6-12 cm long, (1-)1.5-2.7 cm wide, base cuneate to more or less rounded, apex more or less acute, apiculate; upper surface glabrous, lower surface with a rufous, patchy, floccose indumentum over a glaucous, papillate surface. Petiole 7-15 mm long, floccose, usually eglandular. Inflorescence a terminal umbel of 4-7 flowers; rachis 2-3 mm long. Pedicels about 10 mm long, tomentose, eglandular. Calyx 1-4(-8) mm long, with 5 caducous rounded lobes, sparsely tomentose or glabrous; margins ciliate. Corolla 3-4 cm long, fleshy, tubular-campanulate, usually crimson-scarlet with deep crimson nectar-pouches at base, sometimes pink or red and yellow, orange, yellowish or pink, or the tube greenish within. Stamens 10; filaments glabrous. Ovary tapering into the style, densely stellate-tomentose with a lobulate, glabrous basal disk. Style glabrous for most of its length, tomentose at extreme base. Stigma barely exceeding style in diameter. Capsule 1-2.5 cm long, straight to curved, rufous-tomentose.

Distribution. China (north-west Yunnan, south-east Xizang).

Habitat. Common in open pine forest or scrub, or on limestone cliffs in crevices and among rocks, 2700-4000 m.

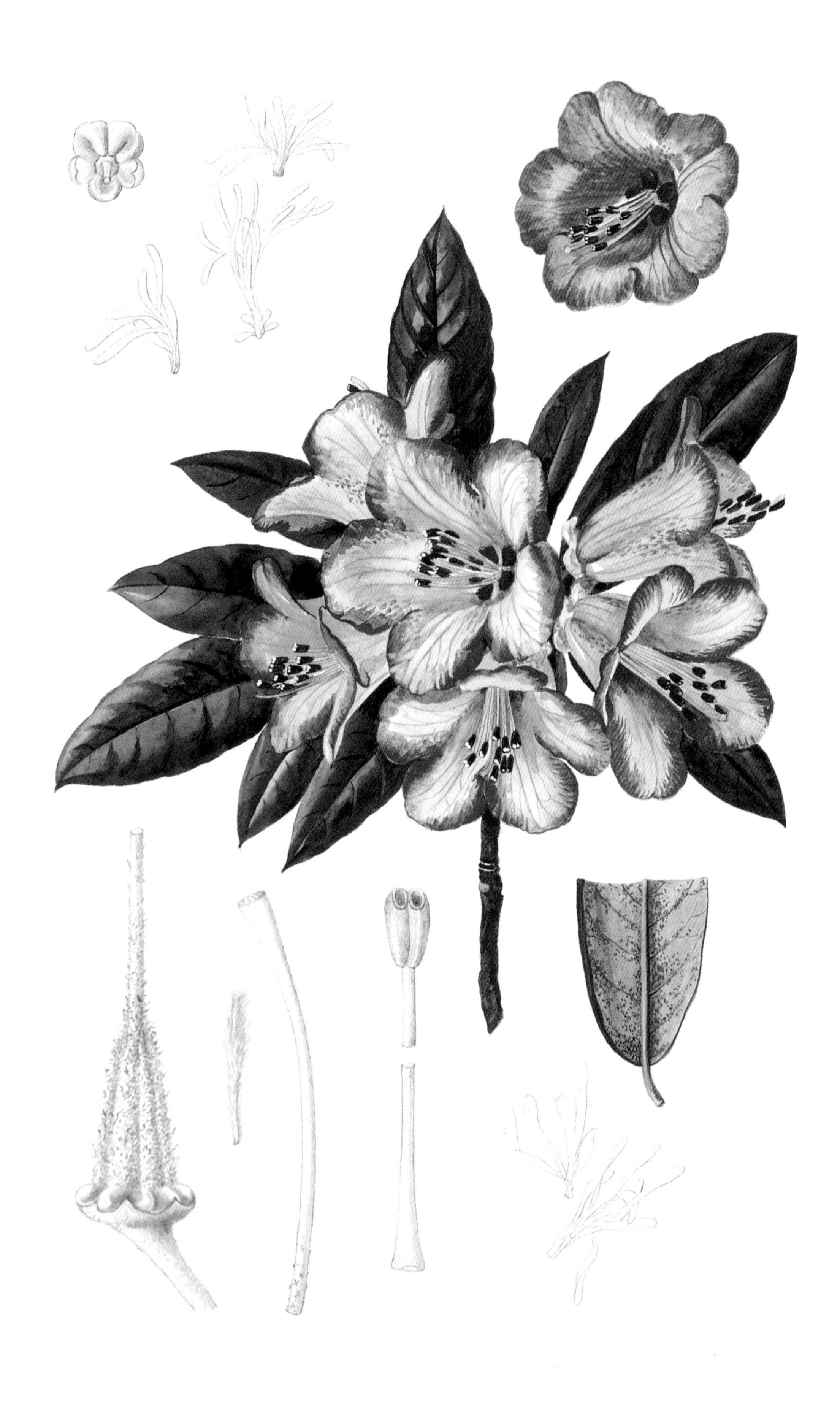

Rhododendron floccigerum
LILIAN SNELLING

RHODODENDRON FULGENS

Subgenus *Hymenanthes*
Section *Ponticum* subsection *Fulgensia*

The flowers of this species are the richest red, perhaps, of the whole genus. From February to April they are produced in small compact trusses of 8-14 which are enclosed in crimson floral bud-scales, an arrangement which serves to concentrate the colour. It also emphasises the contrast between the flowers and the pale, clear viridian green of the young leaves. Birds find the flowers particularly attractive, and peck holes in the darker coloured nectar-pouches at the base of each corolla and, presumably, consume the nectar within. The under surface of the mature leaves is covered with an attractive, usually thick, tawny felt, although both the colour and the density of the indumentum vary somewhat.

Rhododendron fulgens is not considered to be entirely hardy. Certainly it should be given protection from wind, but even so, it is easy to grow in most British gardens. It is a very handsome plant and a worthy recipient of the Award of Merit which it received in March 1933 from the RHS, when exhibited by Gerald Loder of Wakehurst Place, Sussex.

This species was first discovered by J.D. Hooker in the Sikkim Himalaya. He described it in 1849 and introduced it into cultivation at Kew in 1850. The accompanying plate was drawn by Fitch from a plant of this introduction which flowered at Kew in April 1862. It is interesting to note that over 100 years later, plants from Hooker's original introduction were still surviving at Lochinch (Cox, 1979). *Rhododendron fulgens* proved to be widely distributed in the wild and consequently there were many re-introductions subsequent to Hooker's, by various collectors and from numerous different locations.

The subsection *Fulgensia* to which this species belongs, is related to subsections *Neriiflora*, *Thomsonia* and *Barbata* with which it shares characteristic red tubular-campanulate flowers with nectar-pouches.

Rhododendron fulgens Hook. fil., Rhodo. Sikkim Himal. 27, tab.25 (1849).

Description. An evergreen shrub up to 4.5 m in height with pinkish grey, peeling bark; young growth glabrous. Leaves broadly ovate or oblong-obovate, (7-)9-11(-13) cm long, (4-)5-7 cm wide, base cordate to rounded, apex rounded, apiculate; upper surface glossy dark green and glabrous, lower surface covered with a unistrate, dense, white, fawn, tawny or reddish brown wool or felt, the hairs of which are fasciculate; young leaves glaucous. Petiole 1-2(-2.8) cm long, finally glabrous. Inflorescence a compact, rounded, 8-14-flowered truss; floral bud-scales crimson; rachis 8-25 mm long. Pedicels stained crimson, 7-10(-15) cm long, glabrous. Calyx 1-2 mm long, oblique, stained crimson, shallowly lobed, glabrous. Corolla fleshy, scarlet to blood-red with dark nectar-pouches, not spotted, tubular-campanulate, 2-3.5 cm long. Stamens 10, subequal, included, 1.3-1.8(-2.2) cm long; filaments usually white, glabrous. Ovary glabrous. Style pink to crimson, glabrous. Stigma small, usually black, truncate. Capsule cylindrical, 1.3-3 cm long, purplish, glaucous, curved.

Distribution. East Nepal, north-east India (Arunachal Pradesh, Sikkim, Bengal) Bhutan, China (south-east Xizang).

Habitat. Mixed forest, 3200-4300 m.

Rhododendron fulgens
WALTER HOOD FITCH

RHODODENDRON SHERRIFFII

Subgenus *Hymenanthes*
Section *Ponticum* subsection *Fulgensia*

The classification of this species has proved problematical. Authors agree that it is closely related to *R. fulgens*, but opinions differ as to which subsection it should be referred to. It was first placed in series *Campanulatum* and later its affinity to series *Thomsonii* was stressed. In Chamberlain's revision (1982) it was included as a member of subsection *Fulgensia*, there stated to be intermediate between subsection *Neriiflora* and subsections *Thomsonia* and *Barbata*. Chamberlain notes that 'the fulvous lanate leaf indumentum of *R. sherriffii* effectively excludes it from subsection *Thomsonia* and suggests an affinity with *R. fulgens*'.

The species was described by Cowan in 1937 after Ludlow and Sherriff discovered the plant in 1936, growing in rhododendron forest near Lung in the Tibetan Himalaya, at 3300 to 3600 m. They introduced the species into cultivation that same year (L & S 2751). Plants were raised from this seed by J.B. Stevenson at Tower Court from whence seedlings were sent to the Royal Gardens at Windsor, and it was material from one of those flowering in mid-March which was sent to Kew by Sir Eric Savill and was depicted by Margaret Stones in *Curtis's Botanical Magazine* new series tab. 337. The Crown Estate Commissioners exhibited the species at an RHS show in early March 1966, when it received an Award of Merit.

Rhododendron sherriffii is not often seen in cultivation, but is a worthy garden subject if given protection from frost. It is an interesting plant as well as being a restrainedly handsome species. It has a thick, soft, more or less floccose indumentum on the under surface of the leaves and nodding fleshy crimson flowers produced between February and April, the tubes of which are rather long in relation to the corolla-lobes and also in comparison with the other two members of its subsection, *R. fulgens* and *R. miniatum*.

Rhododendron sherriffii Cowan in Notes Roy. Bot. Gard. Edinburgh 19: 231 (1937).

Description. A large evergreen shrub or small tree 4.5-6(-7) m in height with smooth, red-brown to grey peeling bark and slender branches; young growth stipitate-glandular and mealy. Leaves coriaceous, elliptic or oblong to broadly obovate, 4.5-7.5 cm long, 2.5-4 cm wide, with rounded base and rounded, minutely apiculate apex, shiny dark green and glabrous above when mature, with a dense, soft, fulvous to brown, floccose indumentum beneath. Petioles 1.2-1.5 cm long, finally glabrous. Inflorescence a lax racemose cluster of 4-5(-6) nodding flowers; rachis 3-4 mm long. Pedicels flushed crimson, 1-2 cm long, glabrous. Calyx crimson, 3-6 mm long, with 5, often irregular, rounded, glabrous lobes. Corolla fleshy, rich deep crimson or carmine with darker nectar-pouches, tubular or narrowly funnel-campanulate, 3.5-4 cm long, the lobes hardly spreading. Stamens 10, unequal; filaments glabrous. Ovary glabrous. Style crimson, exserted, slender, glabrous. Stigma red, small, capitate. Capsule about 1.3 cm long.

Distribution. China (South Xizang).

Habitat. Steep hillsides in rhododendron forest above the bamboo line, near the edge of the conifer zone, at altitudes of about 4000 m.

Rhododendron sherriffii
MARGARET STONES

RHODODENDRON THOMSONII subsp. THOMSONII

Subgenus *Hymenanthes*
Section *Ponticum* subsection *Thomsonia*

Introduced by J.D. Hooker from Sikkim in 1850, this familiar and extremely handsome species is native to the Himalaya where it is abundant at altitudes between 3000 and 4000 m. The species first flowered in cultivation at the Stanwell Nurseries, Bonnington Road, Edinburgh in 1857, where it had been grafted onto *R. ponticum*, and it was from this plant that material was sent to Kew by Mr Methuen, to be figured for *Curtis's Botanical Magazine* by Fitch.

Like so many rhododendrons, the species is not completely hardy in British gardens and is too large a plant for many of them, but it does well enough in woodland or in similar sheltered positions. It sets plentiful fertile seed from which plants are readily raised although they take a goodly number of years to reach a free-flowering state. As with other early-flowering species, frost damage is a hazard to *R. thomsonii* which blooms from (March-) April to May, but this is a minor problem, easily dismissed by the spectacle of a healthy plant covered with waxy red flowers, their colour enhanced by the pallid, glaucous green young foliage.

The species received an Award of Garden Merit in 1925 and an Award of Merit when exhibited at an RHS show in 1973 by the Crown Estate Commissioners, Windsor. Some fine hybrids have been produced using *R. thomsonii* as a parent, including 'Ascot Brilliant', the result of a cross with a hardy hybrid which had been raised by Standish in the early 1800s, and many of the more recent grexes have become popular with enthusiasts. Not all the hybrids occur in cultivation, as both subspecies of *R. thomsonii* are known to cross with *R. campylocarpum* in the wild.

Rhododendron thomsonii is the type species of subsection *Thomsonia*. Specifically it is closely related to *R. viscidifolium* which has similar, relatively broad leaves with strongly papillate under surfaces. The subsection includes Tagg's series *Thomsonii* subseries *Thomsonii* and Cullen and Davidian's subseries *Cerasinum*. As a whole, it is probably most closely allied to subsection *Neriiflora*. In Chamberlain's revision (1982) *R. thomsonii* comprises two subspecies, distinguished by geographical range and relative sizes of plant, leaves and calyx, subspecies *lopsangianum* being the smaller of the two in these particulars and native to south Xizang; it also is found over a wider altitudinal range (2500-4300 m) than subsp. *thomsonii* which is restricted to between 3000 and 4000 m.

Rhododendron thomsonii Hook. fil., Rhodo. Sikkim Himal. t.12 (1851) subsp. **thomsonii**.

Description. An evergreen shrub or small tree up to 3.5(-7) m in height with colourful, smooth, peeling bark; young growth sometimes sparsely glandular, but more usually glabrous. Leaves coriaceous, orbicular-obovate or elliptic, 3-7.5(-11) cm long, 2-5.5(-7.5) cm wide, with rounded to cordate base and rounded, minutely apiculate apex, upper surface dark green, glabrous, lower surface pale glaucous green and papillate with some red stipitate glands. Petioles 0.5-2.5(-3) cm long, glabrous or sparsely glandular. Inflorescence a dense, compact truss of 3-12 flowers; rachis 5(-18) mm long. Pedicels 10-27 mm long, usually glabrous, rarely glandular. Calyx green or variably stained red, large, leafy, irregularly lobed, usually cupular, 6-20 mm long, glabrous. Corolla waxy, dark red or crimson, occasionally flecked with darker, blackish red, campanulate, 3.5-5(-6) cm long. Stamens 10, unequal, 2-4.5 cm long; filaments glabrous. Ovary glabrous or glandular with lobulate, glabrous basal disk. Style glabrous. Capsule 1.5-2.5 cm long, calyx persistent.

Distribution. East Nepal, north-east India (Arunachal Pradesh, Sikkim) and Bhutan.

Habitat. *Rhododendron* scrub, coniferous forest, 3000-4000 m.

Rhododendron thomsonii subsp. *thomsonii*
WALTER HOOD FITCH

RHODODENDRON STEWARTIANUM

Subgenus *Hymenanthes*
Section *Ponticum* subsection *Thomsonia*

Rhododendron stewartianum is altogether an attractive plant. Its slender twigs form an open, graceful bush at most some 2 m in height and the remarkable variation in the flower colour, makes it an interesting subject for all but the coldest gardens in Britain. A reasonably hardy species, it unfortunately starts into growth too early in the season to escape frost damage, often producing flowers in February and continuing to bloom until April at the latest.

This is one of many species which should be purchased while in bloom, so that one can be sure of obtaining a vigorous plant of a good colour form. There are many to choose from, for the colour ranges from white through cream to yellow and from pale pink to red and crimson, and there are bicolors with corolla margins of a deeper or different colour as well as the occasional spotted form. In the wild, according to Reginald Farrer, all these colours can be found in one colony of the species.

Rhododendron stewartianum was discovered by Forrest in 1904 in Yunnan, but was first introduced by Farrer and Cox, and also by Kingdon Ward, in 1919, both introductions coming from the same general area; Farrer and Cox's collection was from Chimili and Kingdon Ward's from Imaw Bum, a few kilometres to the west. Subsequently, in the 1920's, Forrest made several gatherings, mostly from localities in the north, but also from Chimili.

The species is most closely related to *R. eurysiphon* and *R. cyanocarpum*. It differs from the former of these two species in having a 3-7-flowered inflorescence (*R. eurysiphon* has only 1-3 flowers in a truss) and from the latter in the number of flowers in the inflorescence, as well as in its densely glandular ovary (*R. cyanocarpum* has 6-11 flowers in a truss and a glabrous ovary). Of these three species, only *R. stewartianum* has a more or less persistent indumentum clothing the lower leaf surface. The accompanying plate by Lilian Snelling, for some unexplained reason, was never published in *Curtis's Botanical Magazine*, but it is included here because it shows some of the flower colour variation within the species. The habit drawing was painted from a plant grown by Mr Armitage Moore, and the separate corollas were painted from material supplied by Major Lionel de Rothschild.

The species first flowered in cultivation at Exbury, Logan and Bodnant in 1930 and a plant exhibited at the RHS from Exbury by L. de Rothschild four years later, received an Award of Merit.

Rhododendron stewartianum Diels in Notes Roy. Bot. Gard. Edinburgh 5: 211 (1912).
R. nipholobum Balf. fil. & Forr. in Notes Roy. Bot. Gard. Edinburgh 13: 277 (1922).
R. aiolosalpinx Balf. fil. & Forr., *loc. cit.*
R. stewartianum Diels var. *aiolosalpinx* (Balf. fil. & Forr.) Cowan & Davidian in Rhododendron Year Book 6: 177 (1951).
R. stewartianum Diels var. *tantulum* Cowan & Davidian, *loc. cit.*

Description. An evergreen shrub up to 2.5 m in height with fawn, peeling bark and slender branches; young growth often glandular. Leaves obovate to elliptic, 4-12 cm long, 2-6.5 cm wide, with rounded base and rounded apiculate apex, upper surface bright green, glabrous, lower surface papillose with some glands and a thin, more or less persistent, fawn indumentum, or becoming glabrous. Petioles 5-20 mm long, sometimes winged, usually glabrous. Inflorescence a loose truss of 2-7 flowers; rachis 5 mm long or less. Pedicels 8-25 mm long, glabrous or sparsely glandular. Calyx cup-shaped, (2-)5-15 mm long, with unequal glandular-ciliate or glabrous lobes. Corolla white or cream to yellow, pale to deep pink, occasionally crimson, sometimes purple-spotted, campanulate to tubular-campanulate, 3.5-5.5 cm long. Stamens 10, unequal; filaments densely pubescent at base. Ovary usually densely glandular. Style glabrous. Capsule 1.5-3 cm long.

Distribution. China (north-west Yunnan, south-east Xizang), north-east Upper Burma.

Habitat. Cane brakes, rocky slopes and hillsides, along streams, sometimes in mixed forest, 3000-4250 m.

Rhododendron stewartianum
LILIAN SNELLING

RHODODENDRON LEPTOTHRIUM

Subgenus *Azaleastrum*
Section *Azaleastrum*

This species is included as a representative of section *Azaleastrum* and as a contrast to *R. ovatum* which belongs to the same subsection. The section is comprised of five species:- *R. hongkongense*, *R. ngawchangense*, *R. vialii*, *R. leptothrium* and *R. ovatum*. *Rhododendron leptothrium*, although very closely related to *R. ovatum*, is considered to be botanically distinct from that species.

Besides being easier to grow than its relatives, *R. leptothrium* is a more attractive and somewhat more colourful plant than its allies. It occupies a limited geographical range in China and north-east Upper Burma, where it grows at elevations above the altitudinal limit of *R. ovatum*. Flower colour immediately distinguishes this species from other members of its section, the corollas being rose to deep rose, with oblong rather than rounded lobes. Two further distinctions are the elliptic leaves and the setulose, somewhat longer calyx-lobes. Starting into growth early in the year and flowering in April and May, the plants frequently suffer frost damage and should be provided with adequate protection from the British climate, even in the most favourable localities, if they are to flourish.

George Forrest first introduced *R. leptothrium* into cultivation from one of his 1912 gatherings (no 9341), the year that he discovered the plant growing at elevations of 2100 to 3600 m in the mountains north west of Tengyueh and on the Shweli-Salween divide. Joseph Rock found it in the same area and between them, he and Forrest collected the species several times. In 1917, Forrest made two further gatherings from which the species was described; one was on the Mekong-Yangtze divide to the north (no 12845), the other at Litiping (no 13881), where the plant grows at the generally higher altitudes of 3000-3300 m. Rock again found the plant in the same area as Forrest, then, travelling south, collected it between Chienchuan and Mekong. The plant from which the accompanying plate (tab. 502 new series) was painted by Lilian Snelling in 1928, was raised at Kew from seed (Forrest 9341, received in 1913) and flowered there for the first time in 1927. This was not the first time the species had flowered in cultivation. Plants at both Caerhays Castle, Cornwall and Rowallane, Co. Down flowered, apparently simultaneously, in May 1920.

Rhododendron leptothrium Balf. fil. & Forr. in Notes Roy. Bot. Gard. Edinburgh 11: 84 (1919).
R. australe Balf. fil. & Forr. in Notes Roy. Bot. Gard. Edinburgh 12: 93 (1966).

Description. An evergreen shrub up to 8 m in height with reddish, minutely puberulous young branches, petioles and the upper surface of the leaf midribs. Leaves thinly coriaceous, at first glossy and bronzed, later becoming bright, dark green, narrowly elliptic to lanceolate, 3.5-12 cm long, 1.7-3.6 cm wide, base rounded or cuneate, apex subacute or emarginate, midrib excurrent as a mucro, reticulate. Petioles (5-)10-15 mm long. Inflorescences axillary in upper axils of leafy shoots, in clusters of 2-4 buds, each bud l-flowered. Pedicels 1.5-2.5 cm long, puberulous and glandular-hairy. Calyx leafy, 9 mm long, lobes oblong or ovate, (3.5-)6-8 mm long, (2.5-)3-4 mm wide, variably ciliate with fine and/or glandular hairs . Corolla rose to light magenta or purple with darker speckling, rotate, about 2.5 cm long, up to 5 cm across; tube about 7 mm long, externally minutely puberulous (and sometimes within also) or entirely glabrous; lobes spreading, at least 1.6-1.8 cm long. Stamens 5, unequal, about 1.3-2.3 cm long, filaments puberulous towards base. Ovary subglobose, the upper half usually setose and glandular-hairy. Style crimson, more or less 2.2-2.3 cm long, glabrous. Stigma discoid, lobulate, 2 mm in diameter. Capsule broadly ovoid, 6-8 mm long, verrucose, with persistent calyx.

Distribution. China (north-west Yunnan and adjacent Sichuan and Xizang), north-east Upper Burma.

Habitat. Open scrub, and in both deciduous and coniferous forest, 2100-3600 m.

Rhododendron leptothrium
LILIAN SNELLING

RHODODENDRON OVATUM

Subgenus *Azaleastrum*
Section *Azaleastrum*

In the introduction to their revision of subgenera *Azaleastrum*, *Mumeazalea*, *Candidastrum* and *Therorhodion*, Philipson and Philipson (1986) pointed out that the revisions of sections *Azaleastrum* and *Choniastrum* must still be considered tentative. They placed *R. ovatum* in the first of these sections and included *R. bachii* in the synonymy of that species. The latter species was previously distinguished by its fringed calyx-lobes, but this has proved to be a variable and therefore an unsatisfactory character.

Lindley first described the plant as *Azalea ovata* from material collected and introduced by Robert Fortune in 1844 from Chusan Island south east of Shanghai, off the coast of Chekiang; he also found it in mainland Chekiang, but the material at present in cultivation has probably derived from seeds collected by E.H. Wilson in 1900 and 1907 from plants growing in the Changyang-hsien district of Hubei, at 1500-2100 m. A plant from this batch of seed, raised at Caerhays and presented to Kew in 1927, was figured as *R. bachii* in tab. 9375 of *Curtis's Botanical Magazine*. The illustration of *R. ovatum* reproduced here (tab. 5064) was painted by Fitch. For all its charm, and despite being commercially available, *R. ovatum* is not often seen in gardens, perhaps because it is not the easiest species to grow, or perhaps its lack of popularity is due to the fact that it is not entirely hardy. Slow-growing, and occurring naturally at altitudes below 2400 m, the British climate is not suitable for it, and it is prone to setbacks. Nevertheless, it rewards perseverance, producing bronzed, or sometimes bright red or pink young leaves, followed by white, pink or light mauve, spotted or un-spotted flowers in May and June in cultivation, a month or two earlier in the wild. Even Joseph Hooker, who had seen so many beautiful rhododendrons at close quarters, described *R. ovatum* as 'A very pretty little shrub'.

Rhododendron ovatum (Lindley) Maxim., Rhododendr. As. Orient. 45 (1870) *pro parte*.
Azalea ovata Lindley in J. Hort. Soc. London 1: 149 (1846).
R. bachii Lév. in Feddes Repert. 12: 102 (1913).
R. lamprophyllum Hayata, Icon. Pl. formos. 3: 135 (1913).
R. ovatum (Lindley) Maxim. var. *prismaticum* Tam in Bull. Bot. Res. 2: 99 (1982).

Description. An erect, bushy, evergreen shrub up to 4 m in height, with pale bark and minutely puberulous young growth which is sometimes also glandular-hairy, and often bright pink to red or bronzed. Leaves dark green, glossy, somewhat thin in texture, especially when young, broadly ovate or ovate-elliptic, 3-6 cm long, (1-)1.6-2.6 cm wide, with rounded or cuneate base and acute or obtuse, sometimes long-mucronate, often emarginate apex; upper surface of midrib shortly downy; margins often bearing bristle-tipped teeth, leaves otherwise glabrous. Petioles 6-16 mm long, minutely pubescent, sometimes also glandular-hairy. Inflorescence axillary in clusters near the ends of leafy shoots. Outer floral bud-scales short, inner elongate, 1.2-1.4 cm long, all externally mealy. Pedicels 1.5-2 cm long, puberulous, more or less glandular-hairy. Calyx leafy, lobes oblong or ovate, rounded, 4-7 mm long, 3-4 mm wide, externally mealy, glabrous or ciliate with fine hairs and/or sessile or stalked glandular hairs. Corolla rotate, almost flat, white, pink or pale mauve, upper 3 lobes usually spotted with crimson or purple, (2.5-)4-5 cm across; tube short, minutely pubescent externally and/or within; lobes rounded, spreading, longer than the tube; or corolla glabrous. Stamens 5, unequal; filaments hairy towards the base. Ovary subglobose, about 2.5 mm long, setose and glandular-hairy. Style as long as the longest stamens. Capsule broadly ovoid, about 7 mm long, verrucose, within enlarged persistent calyx.

Distribution. East & central China (Anhui, Zhejiang, Fujian, Guangdong, Jiangxi, Hubei, Sichuan, Guizhou, Guangxi); rare in Taiwan, where it occurs only in the central mountains.

Habitat. Thickets and forests and on open mountainsides, low altitudes, (except in Hubei where it has been found at 2100 m).

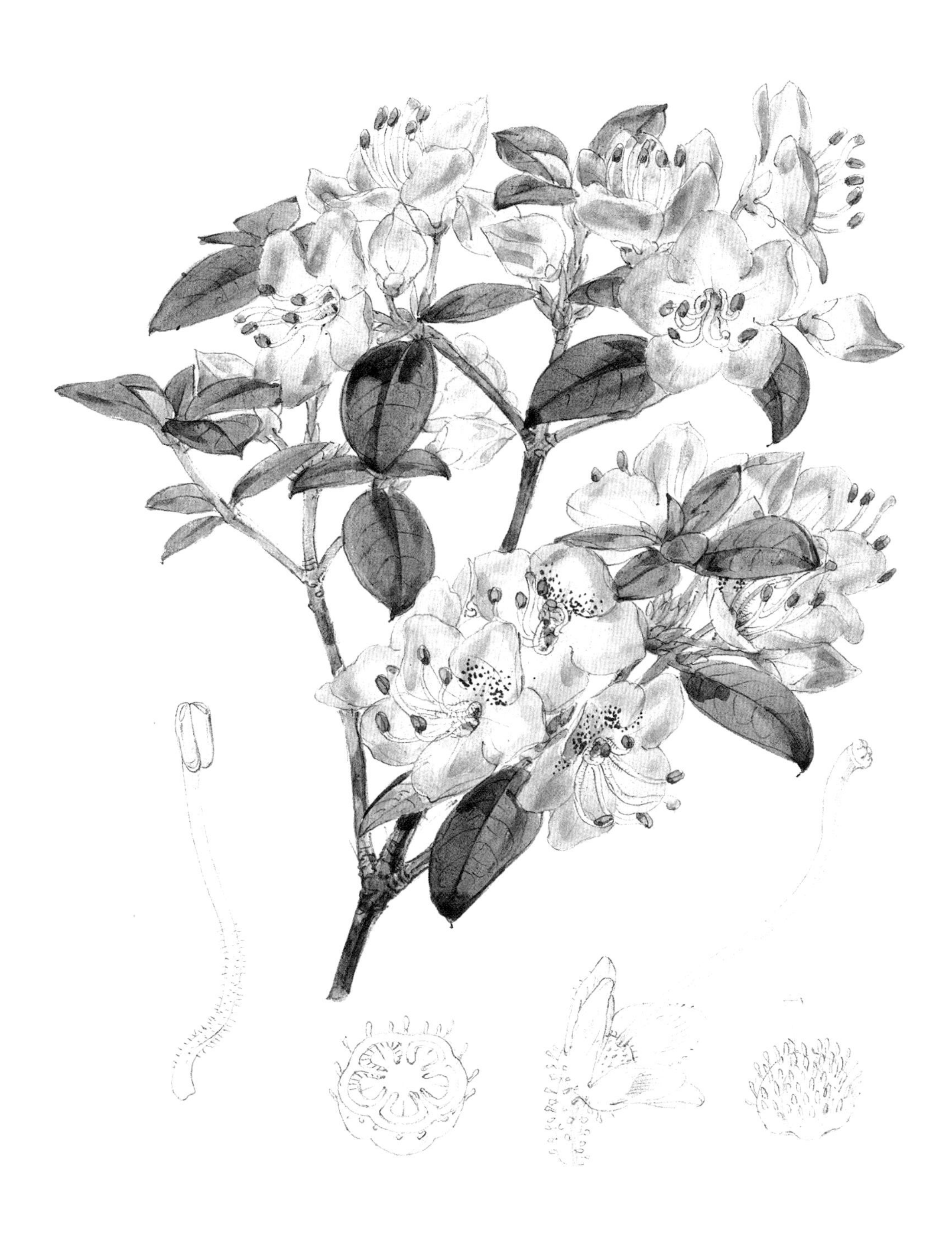

Rhododendron ovatum
WALTER HOOD FITCH

RHODODENDRON MOULMAINENSE

Subgenus *Azaleastrum*
Section *Choniastrum*

This charming rhododendron has hitherto been included in series *Stamineum* in the Bayley Balfour system of classification. Its re-classification is a result of the work of W.R. and M.N. Philipson who revised subgenera *Azaleastrum*, *Mumeazalea*, *Candidastrum* and *Therorhodion* in 1986. In their revision they split subgenus *Azaleastrum* into two sections, *Azaleastrum* and *Choniastrum*, which are distinguished as follows: 'Section *Azaleastrum* differs from section *Choniastrum* in its larger calyx lobes, in having 5, not 10, stamens, the impressed style base and the short ovary and capsule'. In *R. moulmainense* they have included all forms with large funnel-shaped flowers, glabrous pedicels and glabrous leaves, but concede that some of these forms may be sufficiently distinct to be raised to the rank of species in due course.

Rhododendron moulmainense, a wide-ranging species, was first discovered growing in Moulmain on the Gerai mountains, at an altitude of about 1500 m, by Thomas Lobb. Messrs Veitch raised plants from seed in 'a warm greenhouse', and provided the flowering material from which Fitch painted tab. 4904 in January 1856.

In 1917, another species thought to be new, described as *R. stenaulum* by Bayley Balfour and Wright Smith and illustrated by Lilian Snelling, was published in 1944 under tab. 9656. '*Rhododendron.stenaulum*' was discovered by George Forrest in April 1910 in west Yunnan, growing at altitudes between 2100 and 2400 m. Forrest collected other specimens in west Yunnan and he described a new species, *Rhododendron mackenzianum*, in which he also included a specimen collected by Reginald Farrer in the Lang-Yang pass, north-east Burma. Both *R. mackenzianum* and *R. stenaulum* are now considered to be conspecific with *R. moulmainense*.

It is a great pity that this beautiful species is so tender; it is not fully hardy, requiring the protection of a cool glasshouse. The large, fragrant flowers are produced abundantly from March to May and are splendidly offset by the glossy foliage with its pinkish young leaves. The bark, too, is attractive, being much the same texture and colour (pink to red) as that of *Arbutus*. Cox (1979) reported the introduction by Valders of a form with small jonquil-scented flowers in clusters of 16-20, which flowered at Edinburgh Botanic Garden in 1979.

Rhododendron moulmainense received an Award of Merit (under the name *R. stenaulum*) from the RHS in 1937, for a form with silvery lilac, violet-tinged flowers with darker brown-spotted lobes and pale crimson tube.

Rhododendron moulmainense Hook. fil. in Bot. Mag. 82: t. 4904 (1856).
R. ellipticum Maxim. in Bull. Acad. Imp. Sci. St. Pétersbourg 32: 497 (1888).
R. westlandii Hemsley in J. Linn. Soc., Bot. 26: 31 (1889).
R. oxyphyllum Franchet in J. Bot. (Morot) 12: 264 (1898).
R. siamensis Diels in Feddes Repert. 4: 289 (1904).
R. klossii Ridley in J. Fed. Malay States Museum 4: 43 (1909).
R. leucobotrys Ridley, *loc. cit.*
R. leiopodum Hayata, Icon. Pl. formos. 3: 136 (1913).
R. leptosanthum Hayata, *op. cit.* 3: 137.
R. tanakai Hayata, *op. cit.* 4: 15 (1914).
R. stenaulum Balf. fil. & Forr. in Notes Roy. Bot. Gard. Edinburgh 10: 157 (1917).
R. nematocalyx Balf. fil. & W.W. Sm. in Notes Roy. Bot. Gard. Edinburgh 10: 124 (1917).
R. mackenzianum Forr. in Notes Roy. Bot. Gard. Edinburgh 12: 132 (1920).
R. laoticum Dop, Fl. Indochine 3: 735 (1930).
R. pectinatum Hutch. in Gard. Chron. ser. 3, 101: 119 (1937).

Description. Evergreen shrub or small tree to 15 m with glabrous branchlets and leaves. Leaves coriaceous, glossy, crowded towards the tops of the branches, elliptic or narrowly so, 6-17 cm long, 2-5 cm wide, with cuneate base and acute to acuminate apex, young leaves pink-tinged and often ciliate, mature leaves rarely with marginal bristles. Petioles 1-1.8 cm long. Inflorescence axillary, with 4-5(-10) fragrant flowers. Floral bud-scales caducous. Pedicels 1.5-2.5 cm long, glabrous. Calyx obsolete, glabrous; lobes usually minute, pectinate or ciliate but occasionally with 1 or more lobe(s) elongated. Corolla white, pink, lilac or magenta with a yellow blotch in the throat, funnel-shaped with narrow tube 1.6-2.2 cm long and large spreading lobes 3-4 cm long. Stamens 10, unequal, nearly as long as corolla; filaments pubescent towards base. Ovary cylindrical, glabrous. Style exceeding stamens, slender, glabrous. Stigma discoid. Capsule narrowly cylindrical, about 7 cm long.

Distribution. China (south-east Xizang, Yunnan, Guizhou, Guangxi, Hunan, Guangdong, Hainan), Burma, Thailand, Laos, Cambodia, Vietnam, Malaysia (Kedah, Pahang).

Habitat. Diverse situations, but usually in forests, thickets and shady ravines, 900-3000 m.

Rhododendron moulmainense
WALTER HOOD FITCH

RHODODENDRON YEDOENSE var. POUKHANENSE

Subgenus *Tsutsusi*
Section *Tsutsusi*

For many years the double-flowered variety of this plant was believed to be a native of Japan. This fact, together with a plethora of names, caused a great deal of taxonomic and nomenclatural confusion. The details of the case were set out by J.R. Sealy in the text accompanying tab. 455 (new series) of *Curtis's Botanical Magazine* and need not be repeated here. Suffice it to say that Léveillé first used the specific epithet '*poukhanense*' in 1908 when he formally described the Korean Azalea as *R. poukhanense*. It was not until 1920 when Nakai clarified the situation, that the double-flowered variants were referred to *R. yedoense* var. *plenum*, now re-named var. *yedoense* by Chamberlain (1990), The single-flowered plant depicted here, emerged from the confusion with the name *R. yedoense* var. *poukhanense*. It is unfortunate that the misleading epithet '*yedoense*' is the earliest valid name for this Korean plant. The species was formerly included in the Series *Azalea* subseries *Obtusum* but in the Edinburgh revision it has been re-classified as a member of the subgenus and section *Tsutsusi*.

Rhododendron yedoense var. *poukhanense* was first gathered, as far as is known, by W.R. Carles, near Seoul in 1885. The French Missionary Père Urban Faurie collected the same plant on Poukhan, a mountain near Seoul, where Wilson declared it to be rare. On the contrary the plant has proved to be widely distributed and abundant in southern Korea, also occurring on the island of Tsushima. Although *yedoense* does not occur as a wild plant in Japan it is widely cultivated both in that country and Korea. According to Wilson it grows in open country, on grassy mountain slopes and in pinewoods at sea-level and up to about 1600 m altitude, from latitude 38°N southward. The habit varies considerably from low, dense mats to erect loosely branched shrubs about 2 m tall.

The variety *poukhanense* was introduced into cultivation by J.G. Jack, who collected seeds on Mt Poukhan in 1905. Plants were raised from this seed at the Arnold Arboretum, Massachusetts, USA and some of these were sent to Kew in 1913. The variety has been much used in hybridizing which is not surprising, for it is not only reliably hardy, but is also attractive, with large, fragrant, speckled pink or lilac flowers. The slender twigs are brittle and therefore prone to damage, but this is a minor problem. The main difficulty experienced by growers is propagation by cuttings for this variety shows a decided reluctance to root. Although the leaves are usually deciduous, given a mild winter the plant will sometimes retain most of its leaves, shedding them only when the weather becomes colder. On these occasions, the orange and crimson autumn colouring of the foliage is an unexpected bonus. The double-flowered clone 'Yodogawa' received an Award of Merit from the RHS when exhibited in 1961 by Captain Collingwood Ingram. The plant from which the accompanying plate (tab. 455 new series) was prepared by Margaret Stones in April 1960, was raised from seed collected by Mr Moorcraft in 1951 while he was serving with the British Armed Forces in Korea.

Rhododendron yedoense Maxim. in Gartenflora 35: 565, t. 1233, fig. a, b (1886) var. **poukhanense** (Lév.) Nakai in Bot. Mag. (Tokyo) 34: 274 (1920).
R. poukhanense Lév., in Repert. Spec. Nov. Regni Veg. 5: 100 (1908).
R. hallaisanense Lév. in Repert. Spec. Nov. Regni Veg. 12: 101 (1913).
R. yedoense Maxim. var. *hallaisanense* (Lév.) Yamazaki in J. Jap. Bot. 62: 260 (1987).
R coreanum Rehder in Mitt. Deutsch. Dendrol. Ges. 22: 259 (1913).

Description. A dense, twiggy, usually compact, variably deciduous shrub up to 2 m in height with slender, brittle branches, densely covered with adpressed, flattened, red-brown to grey setae when young, eventually glabrescent. Leaves dimorphic, chartaceous, in pseudowhorls of (4-)5(-7) at branch ends and on leafy shoots arising from buds just below the flowers. Spring leaves elliptic to oblanceolate (2.5-)3(-8) cm long, 1-2.5 cm wide with cuneate base and acute, mucronate apex, margins entire, ciliate, bright green above, paler green beneath, sparsely adpressed-setose on both surfaces; summer leaves differ in being of thicker texture with the upper surface soon becoming glabrescent. Petioles 3-6 mm long, covered with adpressed red-brown to grey setae. Inflorescence terminal with 2 or 3 fragrant flowers, opening with or just before the leaves. Pedicels 5-10(-11) mm long, densely brown-setose. Calyx green, (5-)8 mm long, 5-lobed, the lobes unequal, oblong to ovate, acute, rounded or incised at apex, densely covered externally with brown or straw-coloured, flattened setae. Corolla rose-pink to lilac, the upper 3 lobes speckled with scarlet to crimson, widely funnel-form when fully expanded, sometimes double in cultivated plants, (2.8-)3.5-4 cm long, (4-)5 cm in diameter; tube widening upwards, 1.3-1.8(-2.2) cm long, glabrous; lobes somewhat unequal, broadly ovate to ovate-orbicular, sometimes erosulose, (1-)1.5-2.2(-2.6) cm long, or the uppermost shorter, 0.9-1.9(-2.2) cm wide. Stamens 10, subequal, shorter than the corolla; filaments pink, papillose below the middle. Ovary densely adpressed-setose. Style pink, 2.6-4.5 cm long, glabrous or pilose in the lower part. Stigma red-brown, capitate. Capsule ovoid, 5-8(-10) mm long, densely adpressed-setose, in persistent calyx.

Distribution. Var. *poukhanense* is restricted to Korea and the island of Tsushima. Var. *yedoense* is known only in cultivation.

Habitat. Rocky and grassy mountain slopes among shrubs and in thickets and pinewoods, sea-level to about 1100 m.

Rhododendron yedoense var. *poukhanense*
MARGARET STONES

RHODODENDRON SIMSII var. SIMSII

Subgenus *Tsutsusi*
Section *Tsutsusi*

The confusion surrounding the specific names '*indicum*' and '*simsii*' is familiar to all growers and many potential customers at garden centres and plant nurseries have been made uncomfortably aware of the muddle involving these rhododendrons when trying to obtain one or the other species. Chamberlain's revision of the group, published in 1990, has clarified the situation, hopefully to everyone's satisfaction. It should be noted that the distribution of *R. simsii*, the so-called 'Indian azalea', although admittedly wide, does not include India.

Previously a member of series *Azalea* subseries *Obtusum*, *R. simsii* is now re-classified as a member of subgenus *Tsutsusi* section *Tsutsusi* which embraces subseries *Obtusum* and subseries *Tashiroi* in their entirety. Chamberlain (1990) divided this species into two varieties, var. *simsii* and var. *mesembrinum*, which he distinguished as follows:-

Corolla rich red to carmine, 35-60 mm long . . . . var. *simsii*
Corolla white to rose-pink, 25-40 mm long . . . . var. *mesembrinum*

Rhododendron simsii is the main ancestor of the indoor 'azalea' hybrids and cultivars, readily available in the florist's trade, which are such popular house plants. Many are forced into flower for the Christmas market but it is always worth trying these plants in the garden after they have finished flowering. Until such time as the weather becomes mild enough to plant them out, they should be given a cool, moist atmosphere to prevent them dropping their leaves, and the roots should not be allowed to dry out. Although these plants are tender, it is possible to grow them successfully out of doors in sheltered situations in the mildest areas of Britain. Even elsewhere, given reasonable conditions, they will provide a gloriously colourful ground cover in May for several years before a wet British winter finishes off any plants which have survived previous colder, frosty seasons ! In the wild the species covers vast areas with a sheet of scarlet blossom, the sight of which has evoked comments likening it to 'an interminable bloodstain' (Farrer) and 'an active volcano at night' (Kingdon Ward). Its display in Britain is somewhat more modest.

Sometime before 1812, *R. simsii* was introduced into Europe from China, where it had long been cultivated. George Forrest collected seed of the species more than once, but mainly from cultivated plants. A plant which was raised from his seed by G.E. Loder at Wakehurst Place in Sussex, received a First Class Certificate from the RHS in 1933.

The plant figured by Sydenham Edwards for *Curtis's Botanical Magazine* in 1812, as *Azalea indica* (tab. 1480) came from the collection of James Vere Esq. At that time, Sims remarked in the accompanying text 'We believe there are not above three or four individuals of it in the country, and of these only [this one] has as yet produced any flowers'.

Rhododendron simsii Planchon in Fl. des Serres 9: 78 (1854) var. **simsii**.
R. indicum Sweet var. *ignescens* Sweet, Brit. Fl. Garden ser. 2, 2: t. 128 (1833).
R. calleryi Planchon Fl. des Serres 9: 81 (1854).
R. indicum Sweet var. *simsii* (Planchon) Maxim., Rhododendr. As. Orient. 38 (1870).
R. indicum Sweet var. *formosanum* Hayata, Icon. Pl. formosan. 3: 134 (1913).
R. annamense Rehder, J. Arnold Arb. 10: 182 (1929).
R. bicolor P.X. Tan, Survey Gen. Rhododendron S. China 101 (1983).
R vibernifolium W.P. Fang, Acta Phytotax. Sin. 21: 469 (1983).

Description. A much-branched, twiggy, low-growing shrub up to about 2.4 m in height; young growth densely covered with adpressed, flattened glossy brown setae. Leaves evergreen or semi-deciduous, chartaceous, dimorphic; spring leaves elliptic to ovate or oblong-elliptic, up to 2.5(-7) cm long and about 2.5 cm wide, with cuneate base and acute to acuminate apex, sparsely strigose on both surfaces, less so above; summer leaves elliptic to oblanceolate, 1-2(-3.8) cm long, 0.5-1 cm wide, more leathery. Petioles up to 6 mm long, densely strigose. Inflorescence a cluster of 2-6 large flowers. Pedicels 5-10 mm long, densely strigose. Calyx 2-5 mm long with 5 ovate to lanceolate, strigose, ciliate lobes. Corolla rose-red to bright or dark red and spotted, (white, pink, red and white, various shades of red to purple in cultivated forms), widely funnel-form, (2.5-)3.8-6.3 cm across, with broad lobes. Stamens usually 10, rarely 8 or 9, more or less equalling corolla in length; filaments pubescent in lower half. Ovary strigose. Style exceeding corolla, glabrous or strigose at base. Capsule ovoid, about 8 mm long, setose.

Distribution. Very widely distributed over much of west, central, south and east China, and from north-east Upper Burma, south to Thailand and east to Hong Kong, Taiwan and the Ryukyu Islands off south Japan.

Habitat. Forests and scrub on hillsides, cliffs and river banks — sometimes in places where it is submerged during the rainy season, 300-2400 m.

Rhododendron simsii var. *simsii*
SYDENHAM EDWARDS

RHODODENDRON RETICULATUM

Subgenus *Tsutsusi*
Section *Brachycalyx*

In the revision of subgenus *Tsutsusi* which has recently been published by Chamberlain (1990), *R. reticulatum* is placed in section *Brachycalyx*. Chamberlain notes that this species apparently hybridizes with *R. tashiroi*. The section comprises 15 species and includes part of the old series *Azalea* subseries *Schlippenbachii* to which *R. reticulatum* previously belonged.

Rhododendron reticulatum was originally described by D. Don in 1834 from a young plant growing in Knight's nursery in Chelsea that was not in flower. For many years it was known by the later name *R. rhombicum* Miquel, published in 1860 but this name is now reduced to synonymy under *R. reticulatum*. The Japanese recognise a number of closely allied species, all of which are restricted to Japan. The status of some of these is uncertain. *Rhododendron reticulatum* was re-introduced to Europe in the mid-1860's whence it soon reached Britain. It does best in the southern counties of England. The flowers are produced from April to June and are large, showy and precocious, but the flower colour, in various shades of lilac, magenta or purple, or occasionally white, is said to be unpopular.

Nevertheless, if the flower colour fails to please, the species is worth growing for the autumn colouration of the foliage, which is rich, ranging from yellow and red to crimson and purple. Another fault, if such it can be called, is that the species takes some years to bloom, and as the plant flowers before the leaves appear, it cannot be considered an attractive sight in early spring when the branches are bare even though colour in the garden is at a premium.

The accompanying plate (tab. 6972) was prepared by Matilda Smith from a plant flowering in the Azalea bed at Kew in 1887.

Rhododendron reticulatum D. Don in G. Don, Gen. Syst. 3: 846 (1834).
R. rhombicum Miq. in Ann. Mus. Bot. Lugdano-Batavum 2: 164 (1865).
R. rhombicum Miq. var. *albiflorum* Makino in Bot. Mag. (Tokyo) 18: 66 (1904).
R. sakawanum Makino in J. Jap. Bot. 3: 11 (1926).
R. reticulatum D. Don. var. *parvifolium* Yamazaki in J. Jap. Bot. 59: 209 (1984).
R. reticulatum D. Don. var. *bifolium* Yamazaki in J. Jap. Bot. 62: 288 (1987).

Description. A lax, deciduous, erect, much-branched shrub or bushy tree up to 8 m in height; young growth more or less brown-hairy, soon glabrous. Leaves membranous, dark dull green above, paler beneath, 2 or 3 at the ends of branches, rhombic-ovate, sometimes stained red, 3-6(-6.3) cm long, 1.5-4(-5.5) cm wide, with cuneate base and acute, mucronulate apex, upper surface short-hairy, soon glabrous, lower surface shortly brown-hairy, especially on midrib and veins, reticulation conspicuous. Petioles 2-5 mm long, flattened, usually brown-hairy. Inflorescence 1- or 2-flowered, rarely 3- or 4-flowered, flowers precocious. Pedicels more or less 7 mm long, with adpressed brown hairs, glandular. Calyx small, minutely toothed, ciliate, villous or glabrescent, glandular. Corolla pink, lilac, magenta or purple, rarely white, usually without spots, funnel-campanulate, 2.5-3(-5) cm long, somewhat zygomorphic, with short, wide tube and spreading oblong lobes, the upper 3 lobes erect, the lower 2 lobes longer and divergent. Stamens 10, unequal; filaments glabrous. Ovary densely villous. Style exceeding stamens, glabrous. Capsule cylindrical, curved, ribbed.

Distribution. Japan. (south Honshu, Shikoku, Kyushyu).

Habitat. Forested hillsides, 400-700 m.

Rhododendron reticulatum
MATILDA SMITH

RHODODENDRON CALENDULACEUM

Subgenus *Pentanthera*
Section *Pentanthera* subsection *Pentanthera*

This well-known 'Azalea' is a native of eastern North America and one of the most colourful species of that group, a feature which has earned it the names of 'flame azalea' and 'fiery azalea'. It is deservedly a popular garden subject, not only because of its bright orange red, or sometimes yellow flowers, but also because it is completely hardy, flourishing in Britain despite the rigours of climate. In the wild it covers whole hillsides and when in full flower during April, May and early June, the red glow strongly resembles a fire on the slopes. The autumn colouring of this species is almost as spectacular, covering a range from orange to crimson. Not surprisingly the species has been much used in hybridizing and many red or orange garden hybrids owe their colour to this species. It is a parent of numerous attractive progeny, one of which, the clone 'Burning Light', received an Award of Merit from the RHS in 1965 when exhibited by the Crown Estate Commissioners, Windsor.

The discovery of *R. calendulaceum* was probably made by William Bartram in or about 1774, when he found it growing on hillsides in 'West Georgia and lower Cherokee country' and was impressed by its brilliant colouring. As to the date of its introduction into cultivation, it is possible that the species may have been introduced into England before 1806 when John Lyon brought material over from America, as *Azalea aurantiaca*. The latter species, described by Dietrich in 1803, is now considered to be synonymous with *R. calendulaceum*.

The plate from *Curtis's Botanical Magazine* (tab. 1721), reproduced here, was painted by Sydenham Edwards in June 1812 from a plant supplied by 'Mr Lyon' — presumably the same John Lyon who had introduced the species six years before. A previous plate of var. *flammea* of this species (tab. 2143) was published in *Curtis's Botanical Magazine* in 1820. It was drawn by J. Curtis from material sent by 'Mr Thompson of Mile End, the worthy successor of the late celebrated Mr James Gordon whose Nursery was one of the first that rose to botanical celebrity in this country'.

Rhododendron calendulaceum is placed in subsection *Pentanthera* of section *Pentanthera* by Kron in his 1993 revision. The subsection includes many species which were previously placed in series *Azalea* subseries *Luteum*.

Rhododendron calendulaceum (Michaux) Torrey, Fl. N. Mid. U.S., 1: 425 (1824).
Azalea calendulacea Michaux, Fl. bor.-amer. 1: 151 (1803).
A. speciosa Willd., Berl. Baum. ed. 2, 14 (1811), *nom. illeg.*
R. speciosum (Willd.) Sweet, Hort. brit. ed. 2, 343 (1830), *nom. illeg.*
A. calendulacea var. *crocea* Michaux, Fl. bor.-amer. 1: 151 (1803).
R. calendulaceum f. *croceum* (Michaux) Rehder in Mitt. Deutsch. Dendrol. Ges. 24: 225 (1915).
A. aurantiaca Dietr., Darst. Vorz. Zierpfl. 4, tab. 1 (1803).
R. calendulaceum var. *aurantiacum* (Dietr.) Zabel in Beissner, Schelle & Zabel, Handb. der Laubholz. 380 (1903).
R. calendulaceum f. *aurantiacum* (Dietr.) Rehder, Monogr. Azaleas 130 (1921).
A. coccinea Lodd., Bot. Cab. 7: 624 (1822) as '*A. coccinea* var. *major*'.
A. speciosa var. *coccinea* (Lodd.) DC., Prodr. 7: 717 (1834) *nom. illeg.*
A. crocea Hoffmanns., Verz. Pflanzen. Kult. Suppl. 3: 22 (1826) *nom. nov.*
A. speciosa var. *major* Sweet, Hort. brit. ed. 2, 343 (1830).
A. coccinea var. *major* Lodd., Bot. Cab. 7: 624 (1822) *nom. inval.*
A. speciosa var. *aurantia* Lodd., Bot. Cab. 13: 1255 (1827).

Description. A deciduous, much-branched shrub up to 3(-5) m in height; young growth pubescent and bristly. Leaves membranous, broadly elliptic to elliptic-oblong or obovate-oblong, 3.8-8 cm long, 1.3-3.8 cm wide, with broadly cuneate base and acute, mucronate apex, finely pubescent above and more densely so beneath when young, especially on midrib and veins, developing orange to crimson colouring in the autumn before falling. Petioles 2-5 mm long, pubescent. Inflorescence a terminal cluster of 5-7 flowers, expanding with or just after the leaves, the flowers not or only slightly fragrant. Pedicels 6-13 mm long, hairy and glandular. Calyx 1-4 mm long; lobes ovate to oblong, hairy and glandular-ciliate. Corolla varying from orange to red, occasionally yellow, funnel-shaped, 3.8-6.3 cm long, 4-5 cm in diameter; tube externally pubescent and glandular; lobes with wavy margins, equalling or exceeding tube in length. Stamens 5, nearly 3 times as long as corolla-tube; filaments pubescent in lower half. Ovary densely hairy and glandular. Style more or less equalling stamens in length, hairy or glabrous at base. Capsule ovoid-oblong, 15-18 mm long.

Distribution. Eastern USA, from Pennsylvania to Georgia, Carolina and Virginia.

Habitat. Woods and open hillsides, 500-1500 m.

Rhododendron calendulaceum
SYDENHAM EDWARDS

RHODODENDRON LUTEUM

Subgenus *Pentanthera*
Section *Pentanthera* subsection *Pentanthera*

It seems unlikely that this plant, arguably the most fragrant of the 'azaleas', is unknown to anyone with an interest, however fleeting, in the genus *Rhododendron*. Its bright clouds of precocious, strongly scented yellow blossoms in May and June, always attract a great deal of attention and are a source of admiration, whether for the scent or the spectacle they present. Almost as familiar is the knowledge that the honey from this plant has properties described variously as narcotic, intoxicating, poisonous and medicinal ! Certainly the honey is sold in Turkey for medicinal use. The species is completely hardy and can be relied upon to produce plentiful inflorescences each year, followed in due season by fine autumn colouring, fertile seed and, in undisturbed soil, self-sown seedlings in some quantity; it also suckers freely and has become naturalized in some parts of Britain by this means. A popular plant, it is one of the oldest introductions and still one of the most generally cultivated species, as well as being an important parent of yellow hybrids and a dependable stock to receive grafts of less robust varieties.

Classified as a member of series *Azalea* subseries *Luteum* in Stevenson's *The Species of Rhododendron* (1930), *R. luteum* has recently been placed in subsection *Pentanthera* by K.A. Kron (1993). In his revision Kron described *R. austrinum* as *R. luteum*'s closest relative. *Rhododendron austrinum* can be distinguished by its flower colour, which varies from yellow and orange to red with a pink tube, and its comparatively narrow, hairy fruits; *R. luteum* has yellow flowers (with a darker yellow blotch) and almost glabrous fruits.

Rhododendron luteum was originally introduced from the Caucasus by Pallas in 1792, when he sent plants to the nursery of Lee and Kennedy in Britain. It was introduced again in 1798 by Anthony Hove of Warsaw, this time to Watson's of Islington, and it was from this material that the accompanying plate (tab. 433) was drawn by Sydenham Edwards, the plant having been forced into flower by means of artificial heat.

The species has received only one award from the RHS, an Award of Garden Merit in 1930, which is surprising when one considers its popularity.

Rhododendron luteum Sweet, Hort. brit. (ed. 2), 343 (1830), *non Azalea lutea* Linn. (1753).
Azalea pontica Linn., Sp. Plant. ed. 1, 1: 150 (1753).
Anthodendron ponticum (Linn.) Reichb. in Mossler, Handb. Gewächsk. 1: 309 (1827).
R. ponticum (Linn.) Schreber ex DC., Prodr. 7: 718 (1839) *non* Linn. (1753).
R. flavum G. Don, Gen. Syst. 3: 847 (1834) *nom. illeg.*
Azalea flava Hoffmanns., Verz. Pflanzen Kult. Nachtr. 2: 62 (1826).
A. pontica Linn. var. *autumnalis* C. Koch, Linnaea 17: 281 (1843).
R. flavum G. Don var. *macranthum* Bean, Trees & Shrubs Brit. Isles 2: 357 (1914).
R. luteum Sweet var. *macranthum* Wilson, Monogr. Azaleas 105 (1921).

Description. A much-branched, spreading, often stoloniferous shrub up to 3.6 m in height and as much or more in diameter, young growth glandular-hairy. Leaves deciduous, oblong to lanceolate or oblong-oblanceolate, 5-10 cm long, 1.3-3 cm wide, with cuneate base and acute to obtuse and mucronate apex, margin ciliate and serrulate, both surfaces pubescent and glandular-bristly when young, soon both surfaces adpressed glandular-bristly only. Petioles 3-8 mm long, glandular-pubescent. Inflorescence precocious, a more or less lax, 7-12-flowered truss of strongly fragrant flowers. Pedicels 6-20 mm long, glandular-pubescent, sticky. Calyx 2-6 mm long, with 5 unequal, glandular-pubescent and ciliate lobes. Corolla yellow, funnel-shaped, externally densely glandular-pubescent and sticky, about 3.8 cm long overall; tube narrow, about 2 cm long; lobes 5, spreading, ovate, more or less acute, 4.5-5 cm across. Stamens 5, exserted, as long as or exceeding the corolla; filaments hairy in basal half. Ovary densely hairy. Style exceeding stamens in length and hairy at base. Stigma capitate. Capsule cylindrical, up to 2.5 cm long.

Distribution. Mainly in the Caucasus, but *R. luteum* also occurs spasmodically over a wide area of eastern Europe.

Habitat. Dense or open forests, scrub and grassland, sea-level to 1200 m.

Rhododendron luteum
SYDENHAM EDWARDS

RHODODENDRON OCCIDENTALE

Subgenus *Pentanthera*
Section *Pentanthera* subsection *Pentanthera*

One of the finest 'azaleas' in cultivation, this colourful and accommodating species is hardy, not fussy about soil, flourishes in gardens throughout Britain and is both familiar to and popular with growers. Easy to grow, beautiful and flowering in June and July, it has probably been paid more attention than any other species in the genus *Rhododendron*, both in the wild and as a seed parent in hybridization (few seeds result when it is used as a pollen parent), although Lindley in 1857 considered it to be of 'little value'. *Rhododendron occidentale* is by nature extremely variable both in the colour of its flowers and in its autumn colour; in cultivation the selected clones and hybrids extend this variability to the limit, so that those forms commercially available are legion. Once established, the species is free flowering and flowers within three to four years from sowing, but it takes a few years to achieve its full flowering potential. The best clone available is said to be SM 232 'Leonard Frisbie' (with large richly coloured flowers and overlapping frilly lobes) — named after the American who pioneered the study and collection of clones of *R. occidentale* from the wild. Much work had already been done with the species, Anthony Waterer of Knap Hill and Koster of Boskoop being responsible for establishing a race of late-flowering hybrids, by crossing *R. occidentale* with earlier-flowering azaleas.

A formal description of the species was published by Gray in 1876, but *R. occidentale* was discovered much earlier, in 1827, when Douglas and Hartweg collected specimens in California. It was not introduced until 1851 when William Lobb, collecting for Veitch, sent back seed, also from California. The species occurs to the west of the Rocky Mountains, from Oregon south to California, with isolated patches both north and south of this range, and is the only North American azalea to do so. The plate in *Curtis's Botanical Magazine*, (tab. 5005) was painted by Fitch from a plant raised by Messrs Veitch from Lobb's seed.

Until a new classification was published, *R. occidentale* remained a member of Rehder's subseries *Luteum* (of series *Azalea*). When Kron published his revision in 1993, he placed *R. occidentale* in subsection *Pentanthera*, together with thirteen other species. Most closely related to *R. occidentale* are *R. austrinum* and *R. luteum* from which *R. occidentale* is distinguished by its white flowers; *R.austrinum* is further distinguished by its broader fruits.

Rhododendron occidentale (Torrey & Gray) Gray in Brewer & Watson, Bot. Calif. 1: 458 (1876).
Azalea occidentalis Torrey & A. Gray, Pac. R. R. Rep. 4: 116 (1856).
A. californica Torrey & A. Gray ex Durand in J. Acad. Phil. ser. 2, 3: 94 (1855).
A. nudiflora Linn. var. *ciliata* Kellogg, Proc. Calif. Acad. Sci. 1: 60 (1855).
R. sonomense Greene, Pittonia 2: 172 (1891).
R. occidentale (Torrey & A. Gray) A. Gray var. *sonomense* (Greene) Rehder, Monogr. Azaleas 127 (1921).
R. occidentale (Torrey & A. Gray) A. Gray var. *paludosum* Jepson, Man. Fl. Pl. Calif. 741 (1925).

Description. A much-branched, rounded, deciduous shrub up to 4.6 m in height with somewhat downy young shoots, soon glabrous. Leaves chartaceous, oblanceolate, elliptic or obovate, 1.5-9.6 cm long, 0.5-4 cm wide, base cuneate, apex acute to obtuse, margin wavy or straight, ciliate; both surfaces thickly pubescent, glossy green above, paler and more or less glaucous beneath, becoming yellow, scarlet and crimson in autumn. Petiole 2-5 mm long. Inflorescence a more or less lax truss of 6-12 fragrant flowers, expanding with (or rarely before) the leaves. Pedicels 0.5-1.5 cm long. Calyx-lobes up to 7 mm long, ovate, obtuse, densely ciliate. Corolla most often white with yellow flare on upper lobes and often tinged with pink or red, or nearly all pink, or red, or yellow, the flare yellow to orange or crimson or absent, widely funnel-shaped or tubular-campanulate, the tube gradually widening upwards, externally villous and glandular-hairy; lobes 5 or 6, sometimes more, as long as the tube, flat, frilled, sometimes serrated. Stamens 5 (sometimes 6-10), as long as the corolla, but exserted from the tube; filaments pilose in lower half. Ovary glandular-pilose. Style equalling or exceeding the stamens, pubescent at base, rarely glabrous. Stigma discoid, lobulate. Capsule ovoid-oblong, 1.3-3.2 cm long.

Distribution. North America (west of the Rocky Mountains from Oregon to California, with occasional stands as far north as Washington State and very nearly as far south as Mexico).

Habitat. Diverse habitats on many different soils, sea-level to 2700 m.

Rhododendron occidentale
WALTER HOOD FITCH

RHODODENDRON MOLLE

Subgenus *Pentanthera*
Section *Pentanthera* subsection *Sinensia*

In the system of classification most widely used before the recent Edinburgh revision of the genus *Rhododendron* was published, *R. molle* was included in subseries *Luteum* of series *Azalea*. In the Philipsons' preliminary classification of the 'Azalea' groups (1982) it was placed in subgenus *Pentanthera*. A complete account of this group of species has now been published (1993) in which the exact relationships have been defined by K.A. Kron.

In this revision, Kron split section *Pentanthera* into two subsections: subsection *Sinensia*, of which *R. molle* is the sole member and subsection *Pentanthera* which comprises 14 species. Kron divided *R. molle* into two subspecies; subsp. *molle* from China, with yellow flowers and sparsely hairy fruits, and subsp. *japonicum* from Japan, with yellow to orange-red flowers and more densely hairy fruits. *Rhododendron molle* subsp. *japonicum* was illustrated (as *R. sinensis*) for *Curtis's Botanical Magazine* in 1871 (tab. 5905). Fitch's drawing was made from material supplied by a 'Mr Bull' who grew it in Chelsea where it flowered in March 1870. *Rhododendron molle* is abundant in east China, where it covers the hillsides with shades of yellow and orange from April to June. It was first introduced into Loddiges' nursery in England in 1823/4. Later, in 1845, Robert Fortune collected seeds of the species. Wilson, Forrest and Rock all collected the plant, but Rock's collection from Yunnan (no. 59226) may have been a garden escape, as *R. molle* had long been cultivated in the region, as well as in Japan, where it was, and still is a popular garden plant. Not surprisingly this beautiful species has been introduced into cultivation in Europe several times, but it has always been rare in gardens and it is doubtful whether the true species is still grown. This may be due in part to the variable hardiness of the species, which in turn, is explained by its geographical distribution and wide altitudinal range, factors which naturally produce both hardy and tender forms. Many hybrids of *R. molle* exist; the Exbury strain of Knap Hill Azaleas, bred by Lionel de Rothschild derives in part from Rock's no. 59226, and the species has been used with some success as a parent of Azaleodendron hybrids. Its progeny are vigorous and more varied in flower colour than the parent and as a consequence, attract more attention; this factor too, probably has some bearing on the rarity in gardens of the true species.

Rhododendron molle (Blume) G. Don, Gen. Syst. 3: 846 (1834). subsp. **japonicum** (A. Gray) K.A. Kron in Edinburgh J. Bot. 50(3): 279 (1993).
Azalea mollis Blume, Cat. Gewass. Pl. Buitz. 44 (1823).
A. japonica A. Gray, Mem. Amer. Acad. Arts (n.s.) 6: 400 (1859).
R. japonicum (A.Gray) Valcken., Gartenflora 57: 517 (1908).
R. molle var. *japonicum* (A. Gray) Makino, Illustr. Fl. Japan 252 (1956).
A. mollis (Blume) var. *glabrior* Miquel ex Regel, Gartenflora 16: 289, pl. 536 (1867).
A. sinensis var. *glabrior* (Miquel) Maxim., Index Seminum (Petrop.) 1870.
R. glabrius (Regel) Nakai, Trees and Shrubs Japan 1: 64 (1922).
R. japonicum (A. Gray) Valcken. f. *aureum* Wilson, Monogr. Azaleas 102 (1921).
R. glabrius (Regel) Nakai var. *aureum* (Wilson) Nakai, Trees and Shrubs Japan 1: 66 (1922).
R. japonicum (A. Gray) Valcken. var. *canescens* Sugimoto, J. Geobot. 378 (1972).
R. japonicum (A. Gray) Valcken. f. *canescens* (Sugimoto) Sugimoto, J. Geobot. 22: 52 (1975).
A. sinensis Lodd., Bot. Cab. 9: t. 885 (1824).
R. sinense (Lodd.) Sweet, Brit. Fl. gard. 3: t. 290 (1826).
A. pontica L. var. *sinensis* (Lodd.) Lindley, Edwards' Bot. Reg. 15: t. 1253 (1829).
R. sinense (Lodd.) Sweet var. *rosea* Ito, Icon. pl. Japon. 17: t. 2 (1913).

Description. A deciduous, erect, sparsely branched shrub up to 1 m or so in height with the young branches villous and often setose, soon glabrous; buds pubescent. Leaves membranous, oblong to oblong-lanceolate, 6.3-15 cm long, 2-5.5 cm wide, with cuneate base, obtuse, mucronulate apex and often revolute, ciliate margins, upper surface pubescent, at least when young, lower surface densely grey-white pubescent, sometimes only the veins pubescent. Petioles 2-6 mm long, pubescent. Inflorescence 5-10-flowered, flowers fragrant, expanding before the leaves. Pedicels 1-2.5(-5) cm long, glabrescent or glabrous. Calyx small with obtuse, sparsely ciliate lobes. Corolla pale greenish yellow to orange with large greenish flare of small speckles, funnel-shaped, 4-5 cm across; tube wide, minutely pubescent externally; lobes spreading, ovate-oblong, exceeding tube. Stamens 5, as long as corolla; filaments pubescent towards base. Ovary hairy. Style slender, exserted, exceeding stamens, glabrous. Stigma discoid. Capsule ovoid-obloid, 1.5-2 cm long, ribbed, sparsely setose and minutely pubescent.

Distribution. Of subsp. *japonicum*: Japan. [Subsp. *molle* occurs in China (Jiangsu south to Guangdong and west to eastern Sichuan and Yunnan where it is possibly not wild, but merely an escape from cultivation].

Habitat. Open situations in thickets, wooded slopes, moors, and on volcanic ash, 100-2100 m.

Rhododendron molle subsp. *japonicum*
WALTER HOOD FITCH

RHODODENDRON CANADENSE

Subgenus *Pentanthera*
Section *Rhodora*

One of the hardiest and prettiest, early, precocious-flowering shrubs, producing its inflorescences in April, it is a pity that *R. canadense* is not more widely grown. It frequently grows in swamps in the wild, so not unexpectedly delights in a moist situation in cultivation. In the autumn there is the secondary attraction of foliage colour. This is not as spectacular in *R. canadense* as in some other azaleas; the leaves of the pink-flowered forms develop purplish tones and those of white-flowered forms become yellowish in colour.

Initially classed in a genus of its own (*Rhodora*), *R. canadense* was later transferred to *Rhododendron* and included in what was previously called subseries *Canadense* of series *Azalea*. This group has not yet been revised (1995). The Philipsons (1982) placed *R. canadense* with *R. vaseyi* in section *Rhodora*, subgenus *Pentanthera*, but listed no synonyms. (see footnote)

Rhododendron canadense was originally introduced to Britain by Sir Joseph Banks in 1767. It is a native of north-east North America, from Labrador to north-east Pennsylvania and New Jersey, west to central New York and is easily distinguished by its two-lipped corolla which is an odd departure from the usual rhododendron flower-shape, consisting of a three-toothed upper lobe and two, deeply divided, narrow lower ones. Flowers are freely produced and the colour ranges from pink to crimson. There is a particularly fine pure white form which, if kept away from pink-flowered plants, will breed true from seed. *Rhododendron canadense* is the most northerly of the azaleas.

Rhododendron canadense (L.) Torrey, Cat. Pl., 151 (in Geol. Surv. N.Y. Assembly, No. 50) 1839.

Description. A deciduous, erect, much-branched, sometimes stoloniferous shrub up to about 1 m in height; branches erect, slender, more or less whorled, puberulous when young. Leaves elliptic-oblong, 1.9-4.5(-5.7) cm long, 0.8-1.9 cm wide, with cuneate base, more or less acute apex and revolute, ciliate margin, dull blue-green and sparsely strigillose above, thinly grey-tomentose and sparsely glandular-hairy beneath, usually also with scattered, fulvous hairs, midrib downy on upper surface. Petioles slender, up to 7 mm long, puberulous and sparsely strigose. Inflorescence terminal, umbellate, 3-6-flowered, flowers opening before the leaves. Pedicels up to 1 cm long, glaucous, puberulous and usually somewhat strigose. Calyx short, unequally lobed, puberulous and fringed with setae. Corolla pink to rosy purple or white, rotate-campanulate, 1.5-2 cm long, 2-lipped, the upper 3 lobes united for most of their length, the 2 lower narrowly oblong lobes deeply divided and divergent, glabrous. Stamens 10, unequal, about as long as corolla; filaments pubescent in the lower third. Ovary pubescent, sparsely glandular and setose-hairy. Style slightly exceeding stamens, glabrous or minutely puberulous at base. Capsule ovoid-oblong, about 1 cm long, setose and finely pubescent.

Distribution. North-eastern North America from Labrador to north-east Pennsylvania and New Jersey and west to central New York, as stated above.

Habitat. The species prefers moist situations and is frequently found in swamps but also in moist woods and on river banks, altitude has apparently not been recorded.

Note. Status confirmed (1996) and following synonyms listed:

Azalea canadensis (L.) Kuntze, Rev. Gen. Pl. 2: 386 (1891).
Hochenwartia canadensis (L.) Crantz, Inst. 2: 469 (1766).
Rhododendron canadense (L.) Torr. forma *albiflorum* (E.L. Rand & Redf.) Rehder.*
R. canadense (L.) Torr. forma *album* Voss (reference not traced).
R. canadense (L.) Torr. forma *viridifolium* Fernald in Monogr. Azaleas: 122 (1921).
R. pulchellum Salisbury, Prodr. 287 (1796).
R. rhodora J.F. Gmel. forma *albiflora* E.L. Rand & Redf., Fl. Mt. Desert Isl.: 127 (1894).
Rhodora canadensis L., Sp. Pl. ed. 2: 561 (1763).

* In the Gray Herbarium card index, the reference given is: Rehder in E.H. Wilson & Rehder Monogr. Azaleas: 122 (1921).

Rhododendron canadense
SYDENHAM EDWARDS

RHODODENDRON NIPPONICUM

Subgenus *Pentanthera*
Section *Viscidula*

Rhododendron nipponicum was first collected by Matsumura in 1883 and described by him in 1899. He found it growing on Mt Tateyama in the Toyama (Etchu) Province of Japan. The plant is apparently confined to the mountains of central Honshu, having not been found anywhere else in Japan, although it is locally common within its range, wherever the country is open. A remarkably un-rhododendron-like species, *R. nipponicum* was another of E.H. Wilson's introductions, this time collected from the hills around Toge, Hamagata (Uzen or Usen) Province at about 1000 m on 20 July 1914 when in flower, and on 23 October of the same year, in fruit. Philipson & Philipson in their preliminary synopsis of subgenus *Pentanthera* (= Kron's section *Pentanthera*), tentatively placed *R. nipponicum* in a section of its own, section *Viscidula*. This is not a new departure, as the species has always merited and been placed in a class of its own, the tubular corolla setting it apart from its congeners. But as yet, section *Viscidula* has not been revised and the status of *R. nipponicum* remains unclear*.

The species flowers from late June to early July and is hardy, although not breathtakingly beautiful at any time, but in autumn it does develop good foliage colour, bright and colourful enough to challenge the best of the azaleas in shades of orange and red. For that reason alone it should be rewarded with a place in the garden. The plant from which tab. 491 was drawn was grown by Sir Edward Bolitho, in his garden at Trengwainton, Penzance, Cornwall, from whence material was sent to Kew for illustration by Lilian Snelling.

Rhododendron nipponicum Matsumura in Bot. Mag. (Tokyo) 13: 17 (1899).

Description. An erect, deciduous shrub 1-2 m in height with rigid, straight, red-brown branches, pilose and glandular-hairy when young, becoming glabrous, and with peeling, shredding bark. Young shoots arising from lateral buds directly below the terminal inflorescence. Leaves thin, membranous, sessile or subsessile, (5-)7.5-11(-18) cm long, 4-6(-9) cm wide, obovate, base cuneate, apex rounded, usually emarginate, margin often stained red, undulate and ciliate; upper surface bright green and somewhat wrinkled, sparsely setulose, lower surface pale green, sparsely setulose, mainly on midrib and veins. Inflorescence a terminal, condensed raceme of (6-)9(-15) nodding flowers; flowers expanding with or after the leaves; rachis about 4 cm long. Pedicels slender, 1.3-2 cm long, densely glandular-hairy. Calyx deeply divided into 5 unequal lobes 3-5 mm long, externally glandular, margins glandular-ciliate. Corolla greenish white, lobes flushed with pink, buds deep cerise, tubular-campanulate, 1.5-2.5 cm long overall; tube 7-8 mm in diameter at base, widening upwards to 11 mm in diameter at mouth; lobes rounded, erect to slightly spreading. Stamens 10, unequal, 1.5-1.9 cm long, included; filaments pubescent near base. Disk lobed, 1.25-1.5 mm in height. Ovary conical, densely stipitate-glandular. Style 1.3-1.4 cm long, thickened towards apex, glabrous. Stigma capitate, lobulate, 2 mm in diameter. Capsule nodding, ellipsoid, 1-1.5 cm long, verrucose.

Distribution. Japan (mountains of central Honshu).

Habitat. Open country on hillsides and in deciduous forests, 900-1300 m.

* status confirmed (1996).

Rhododendron nipponicum
LILIAN SNELLING

RHODODENDRON ALBRECHTII

Subgenus *Pentanthera*
Section *Sciadorhodion*

Dr Michael Albrecht, physician to the Russian Consulate at Hakodate in Yeso, Japan, for whom this strikingly coloured rhododendron was named, discovered the plant in 1860 in the mountains of Volcano Bay, Yeso. It was then in fruit, but when he returned in 1862, he collected it in flower. Since its discovery, the plant has been repeatedly collected in a number of localities, but it was not introduced into cultivation until Prof. Sargent sent seeds to the Arnold Arboretum from Yeso in 1892 (the plants raised there did not do well). The species was re-introduced from Hondo by Wilson in 1914. A plant which was raised from Wilson's seed by G.W. Johnstone of Trewithen, Cornwall, was presented to Kew in 1925. It flowered at Kew the following year when it was painted by Lilian Snelling for *Curtis's Botanical Magazine* (tab. 9207), but for some time, although much sought after, it remained scarce in cultivation. Lionel de Rothschild wrote in 1934, that he had seen no plant of the species over 3 ft in height and referred to it as being 'rare and fastidious'.

This Japanese endemic species is found in subalpine regions, on forest margins and in thickets, usually on slopes where the soil is relatively loose, at altitudes below 2000 m. It forms a small shrub with a lax, open habit, rather few leaves and vivid pink flowers. In the wild the flowers are produced from May to July and are frequently precocious, but in cultivation in Britain, the flowers appear in early spring (March to May) and almost inevitably incur some frost-damage. Fortunately, young growth lost in this way is readily replaced, as the plant itself is winter-hardy and vigorous, so that despite the damage to flower-buds, it can be relied upon to bloom the following year. In *The Species Rhododendron* (1930), *R. albrechtii* was classed as a member of series *Azalea* subseries *Canadense*. When the Philipson's produced their 'preliminary synopsis of the Genus Rhododendron' in 1982, they placed *R. albrechtii* in subgenus *Pentanthera* section *Sciadorhodion*, together with *R. pentaphyllum*, *R. quinquefolium* and *R. schlippenbachii*. However, when Kron published his revision of section *Pentanthera* in 1993, he omitted *R. albrechtii* from it. The position of *R. albrechtii* therefore, has yet to be defined*.

The species has received two awards from the RHS. On 13 April 1943, an Award of Merit was given to a plant shown from Bodnant, and a First Class Certificate was awarded on 1 May 1962, to a clone 'Michael McLaren' exhibited from the same garden.

Rhododendron albrechtii Maxim. in Bull. Acad. Imp. Sci. Saint Pétersbourg 15: 227 (1870).
Azalea albrechtii (Maxim.) Kuntze, Rev. Gen. Pl., 2: 387 (1891).
Rhododendron albrechtii forma *canescens* Sugim. in J. Geobot., 22(4): 52 (1975).

Description. An erect, deciduous shrub of loose, twiggy habit 1-3 m in height with slender branches; young shoots with crisped, sticky hairs, soon glabrous. Leaves membranous, often in whorls of 5 at the ends of the branches, sometimes scattered along the branches, the young foliage bronzed, obovate-oblanceolate, 3.8-11.5 cm long, 1.3-6.3 cm wide, with tapering base, acute, mucronate apex and serrulate margin, the teeth bristle-tipped, upper surface sparsely adpressed-bristly, lower surface grey-tomentose, especially on midrib and veins. Petioles 2 mm long, winged, hairy. Inflorescence 4-5-flowered, flowers opening before or with the leaves. Pedicels 1-2 cm long, clothed with curled glandular hairs. Calyx small, glandular-hairy. Corolla bright rosy purple to deep rose, upper lobes spotted with green, rotate-campanulate, 1-2 cm long, about 3 or 4 cm across, with a short, wide tube and spreading, rounded lobes, glabrous. Stamens 10, unequal, the longest equalling the corolla; filaments glabrous or the lower half pubescent. Ovary hairy and glandular. Style slender, exceeding stamens, glabrous. Capsule borne erect, conoid-ovoid, about 1 cm long, sticky-hairy.

Distribution. Japan, from mid-Honshu north to central Hokkaido, where it is widespread, but not common.

Habitat. Forest margins and thickets in subalpine regions, up to 2000 m.

* status confirmed (1996)

Rhododendron albrechtii
LILIAN SNELLING

RHODODENDRON SEMIBARBATUM

Subgenus *Mumeazalea*

The subgenus *Mumeazalea* of which *R. semibarbatum* is the only representative, is remarkable for its dimorphic stamens, a character which is possessed by no other species of *Rhododendron*. The species, and therefore the subgenus, is confined to mountainous regions of central and southern Japan where it is widely distributed but nowhere common.

Rhododendron semibarbatum was described by Maximowicz from a specimen collected by Tschonoski. It was first introduced to St Petersburg where it was grown in a cool glasshouse, but presumably did not survive. It was not until 1914 that the species was re-introduced by E.H. Wilson, who sent seed to the Arnold Arboretum where plants were raised from it and distributed. One of these plants was sent to Kew where, although it did not flourish for long, it at least provided enough flowering material to be figured for *Curtis's Botanical Magazine* (tab. 9147); the artist was Lilian Snelling. The species at that time had not proved hardy at the Arnold Arboretum and it remains one of the more difficult subjects for successful cultivation, but is worth growing, if only for the autumn colouring of its foliage, which varies from yellow through orange and red to crimson. The flowers are produced in June, and although pretty enough in close-up, are small, scattered and hidden by the leaves.

Rhododendron semibarbatum Maxim. in Bull. Acad. Imp. Sci. St. Pétersbourg, Ser. 3, 15: 230 [Mél. Biol. 7: 338] (1870).
Azalea semibarbata (Maxim.) Kuntze, Revis. Gen. Pl. 2: 387 (1891).
Azaleastrum semibarbatum (Maxim.) Makino in Bot. Mag. (Tokyo) 28: 338 (1914).

Description. A deciduous, bushy, more or less erect shrub up to 2(-3) m in height, with spreading, irregularly whorled branches, puberulous and glandular-hairy, at least when young. Leaves membranous, clustered at the ends of short shoots or spaced along the longer branches, elliptic or ovate, 2-6 cm long, (0.7-)1-2.6 cm wide, with a cuneate or rounded base and subacute, obtuse or apiculate apex, or emarginate and mucronate, margin minutely toothed, the teeth sometimes bristle-tipped. both surfaces of the midrib puberulous, veins ciliate beneath. Petioles 4-12 mm long, puberulous and glandular-hairy. Inflorescences axillary, crowded beneath the terminal tuft of leaves, each bud l-flowered with persistent floral bud-scales. Pedicels 5-8 mm long, pubescent and glandular-hairy. Calyx-lobes rounded, about 2 mm long, externally hairy and margins glandular-ciliate. Corolla white flushed with pink, with red or purplish speckling, rotate, about 2 cm across lobes; tube short and wide; lobes rounded, spreading, sometimes somewhat reflexed. Stamens 5, unequal; upper 2 erect, short, about 7 mm long with filaments densely pilose for most of their length; lower 3 spreading, more or less 12 mm long, almost glabrous or minutely puberulous at base. Ovary subglobose, the upper half setose and densely glandular-hairy. Style slender, shorter than stamens, glabrous. Capsule subglobose, 4 mm long, densely glandular-hairy.

Distribution. Central and southern Japan.

Habitat. Forests and thickets in mountains, at high altitudes.

Rhododendron semibarbatum
LILIAN SNELLING

RHODODENDRON ALBIFLORUM

Subgenus *Candidastrum*

One of the few American representatives of the genus and classed as the only species in the series *Albiflorum*, this species now forms the monotypic subgenus *Candidastrum* in the Philipsons' revision (1986). Geographically, this subgenus is well separated from other New World rhododendrons and is easily recognized by its nodding, cup-shaped flowers, which are scattered along the branches below the terminal tuft of leaves, rather than clustered at the ends of the branches.

The plant was discovered by Drummond in the Rocky Mountains, where it was growing in alpine woodland near the tree-line. It was introduced into cultivation in 1828 when plants were raised by Dr Graham from seeds of Drummond's collection. These flowered in July 1837, and were illustrated for *Curtis's Botanical Magazine* by Fitch (tab. 3670).

In the wild, *R. albiflorum* forms impenetrable thickets in which the low trailing branches are a hazard to walkers and the plant has earned the nickname 'mountain misery' because of this habit. It certainly cannot be said that the species has universal appeal, nor is it often seen in gardens, although being an aberrant species, it does have scientific interest and some growers appreciate its delicacy and un-rhododendron-like appearance. Flowers are not produced in profusion, even when the plant is flowering freely, and the species is difficult to grow. Andrew Harley grew it successfully in Perthshire in the 1920s and '30s, where it seemed to prefer a poor, stony soil which, being in a fairly sunny situation, tended to dry out. Where it does well, *R. albiflorum* is useful as a late-flowering species, the flowers appearing in June and July when many other rhododendrons have finished blooming.

Rhododendron albiflorum Hook., Fl. bor.-amer. 2: 43 (1834).
Cladothamnus campanulatus Greene in Erythraea 3: 65 (1895).
Azaleastrum albiflorum (Hook.) Rydberg in Mem. New York Bot. Gard., 1: 297 (1900).
A. warrenii A. Nelson in Bot. Gaz. (London) 56: 67 (1913).
R. warrenii (A. Nelson) Macbride in Contrib. Gray Herb. n.s. 56: 55 (1918).

Description. An erect, deciduous shrub up to 2 m or so in height, often producing long, trailing branches from the base of the plant, and long fastigiate branches in cultivation; young growth clothed with brown, strigose hairs. Leaves membranous, bright green, scattered along the branches or clustered at branch ends, narrowly elliptic, oblong or lanceolate-oblanceolate, 2.5-8 cm long, 1-2.4 cm wide, with cuneate base, acute apex and minutely toothed margin; young leaves ciliate with adpressed-hairy midrib, finally more or less glabrous. Petioles 1.5-4 mm long, strigose. Inflorescences axillary, scattered along previous year's shoots, each bud 1-2-flowered, flowers nodding, borne below terminal leaf-cluster. Pedicels 5-10 mm long, minutely glandular-hairy and with strigose hairs at base. Calyx deeply divided with oblong-ligulate lobes, 8-13 mm long, 3-4 mm wide, externally strigose and glandular-hairy with glandular-hairy margins. Corolla creamy white, rarely spotted with yellow or orange, bowl- or cup-shaped, about 2 cm across, 5-lobed (occasionally 6-lobed), with a short, broad tube and more or less spreading lobes, usually more or less puberulous on both surfaces. Stamens 10 (occasionally 12), equal; filaments 1 cm long, hairy towards base. Ovary globose, setose and glandular. Style straight, impressed, lower half hairy. Stigma discoid. Capsule globose, shorter than persistent calyx, about 8 mm long, glandular and strigose-hairy.

Distribution. Canada (British Columbia, Alberta), U.S.A. (from Washington and Montana to Oregon and Colorado).

Habitat. Upper montane forests where it forms dense thickets along streams and in open places just above the treeline, 1200-2200 m.

Rhododendron albiflorum
WALTER HOOD FITCH

RHODODENDRON CAMTSCHATICUM

Subgenus *Therorhodion*

Rhododendron camtschaticum has variously been placed in the genera *Therorhodion* and *Rhodothamnus* as well as *Rhododendron* (series *Camtschaticum*). It was removed from *Rhododendron* on account of two characters: the unilaterally split corolla and the habit of producing flowers on the current year's growth, instead of from overwintering buds. It is now replaced in subgenus *Therorhodion* by the Philipsons (1982). A plant with a northerly distribution, this species does well in Scotland and in Scandinavia, producing free-flowering mats of dense, prostrate growth, but it does not flourish in the more southerly parts of Britain. It is considered to be very hardy, yet early frosts can and do damage the young growth sufficiently badly to prevent the plant from flowering wherever it is grown; nor is it easily raised from seed, as germination is frequently poor. Once established however, seedlings will flower within their first year, and mature plants, when growing vigorously and in full bloom, make a most attractive spectacle, creeping over large areas and rooting as they go. *Rhododendron camtschaticum* is a deciduous species and sometimes produces good foliage colour in the autumn, as well as numerous comparatively large rosy purple, spotted flowers in May and June. Pink and red forms are also in cultivation and according to Cox (1985) there is a rare white form which has been recently introduced from Alaska, but this form is difficult to maintain beyond the seedling stage. The species likes plenty of sun, but care should be taken to prevent the roots drying out. In 1900, seed of *R. camtschaticum* was acquired from St Petersburg Botanic Garden. Tab. 8210, published in 1907-08, was painted by Matilda Smith from a plant raised at Kew from that seed.

Rhododendron camtschaticum Pallas, Fl. Ross. 1: 48, t. 33 (probably 1789).
Rhodothamnus camschaticus (Pallas) Lindl., Paxton's Fl. Gard. 1: t. 22 (1853).
Therorhodion camschaticum (Pallas) Small, N. Amer. Fl. 29(1): 45 (1914).
R. camschaticum Pallas subsp. *typicum* Hultén, Fl. Kamtschatka 4 :14 (1930), *nom. inval.*

Description. A deciduous low-growing or prostrate undershrub up to 30 cm in height, usually less, with ascending branches. Leaves chartaceous, 2-5 cm long, 1-2.5 cm wide, with cuneate base and rounded to obtuse apex, midrib, primary veins and margins setose, venation prominent. Inflorescence terminal, flowers solitary or in pairs at ends of the current year's branches. Pedicels 1-1.5 cm long, bearing leafy bracts, setose-glandular. Calyx leafy, deeply divided into 5 oblong lobes, 1-1.8 cm long, 4-5 mm wide, each with 3 prominent veins, externally setose-glandular, ciliate. Corolla usually rosy purple, shallowly funnel-campanulate, 5-lobed, upper lobes usually spotted; tube split almost to base on lower side; lobes oblong-elliptic, rounded, externally pubescent, about 2.5 cm long, 4 cm across. Stamens 10, unequal, 12 cm long; filaments red-purple, densely pubescent towards base. Ovary ovoid, pubescent. Style red-purple, slender, declinate, hairy towards base. Stigma capitate, lobulate. Capsule narrowly oblong-ovoid, about 1 cm long, shortly pubescent.

Distribution. Japan (north Honshu, Hokkaido), Russia (south and east Kamtchatka, Kurile Is., USA (Aleutian Is. and the coast of south Alaska).

Habitat. Gravelly loam and in rock crevices, often on hilltops, altitude, though rarely recorded, is often in excess of 2000 m.

Note. Chamberlain *et al.* (1996) accepted 2 subspecies of *R. camtschaticum*, subsp. *camtschaticum* (syns: *R. kamtschaticum*, *Rhodothamnus camschaticus* (Pall.) Lindl., *Therorhodion camschaticum* (Pall.) Small) and subsp. *glandulosum* (Small) Hultén, Korgl. Svenska Vetenskapsakaal Handl., ser. 3, 8(2): 15 (1930) (syns: *R. glandulosum* (Small) Hutch., *Therorhodion glandulosum* Small).

Rhododendron camtschaticum
MATILDA SMITH

RHODODENDRON × AGASTUM

Subgenus *Hymenanthes*
Section *Ponticum* subsection *Fortunea* × subsection *Arborea*

Rhododendron × *agastum* was discovered by Forrest in May 1913 at the head of the Taping-pu valley (altitude 2100-2400 m) in west Yunnan. He found it again in the Yangpi valley, this time at 2700 m, in April 1914 and made a third collection in April 1917 on the Shunpi-Yangpi divide at 2400 m. Both Rock's and McLaren's collectors also found the plant in the same area. It appears to be confined to a very limited area around Yangbi on the western flank of the Cangshan, a mountain range close to Dali. At first glance some forms of *R.* × *agastum* are strongly reminiscent of *R. irroratum*, though it differs in its 7-lobed corolla. It was therefore wrongly assigned to subsection *Irrorata*. Forrest stated on the label accompanying the type specimen that he thought that the plant was a hybrid between *R. fortunei* (subsection *Fortunea*) and *R. arborea* subsp. *delavayi* (subsection *Arborea*). In fact *R. fortunei* does not occur in western China so he was almost certainly referring to *R. decorum*. Undoubted hybrids between *R. decorum* and *R. arboreum* subsp. *delavayi* do occur close to Yangbi, lending support to Forrest's original diagnosis.

As its name indicates ('*agastum*' means 'charming'), *R.* × *agastum* is an attractive plant, free-flowering, and producing many-flowered trusses of white or pink, usually heavily spotted flowers (although they are sometimes unmarked) from February or March to May. Despite its appeal, and rather surprisingly perhaps, it seems to be rare in cultivation. Coming from comparatively low altitudes in south-west China, this species is not expected to be completely hardy, but it flourishes in milder gardens and will usually succeed in colder areas if given protection while young, because it becomes hardier as it matures. In cultivation this hybrid has been confused with *R. papillatum*, a species from the Indo-Himalayas that is correctly assigned to subsection *Irrorata*. The plate in *Curtis's Botanical Magazine* (tab. 9577) was painted by Lilian Snelling from a plant grown by E.J.P. Magor in his garden at Lamellen, Cornwall, which flowered in early April 1929.

? **Rhododendron decorum** Franchet × **R. arboreum** Sm. subsp. **delavayi** (Franchet) Chamberlain.
R. agastum Balf. f. & W.W.Sm. in Trans. Bot. Soc. Edinburgh, 27: 178 (1917).

Description. An evergreen shrub or small tree up to 3(-6) m in height, with rough bark; young growth thinly floccose and stipitate-glandular. Leaves coriaceous, obovate-elliptic, 6-11(-15) cm long, 2.5-5 cm wide, with rounded base and rounded, acuminate apex, upper surface glabrous, lower surface fawn or brown, thinly covered with dendroid hairs embedded in a surface film and veins with persistent red-punctate hair-bases. Petioles 1.5-2(-3) cm long, glabrous. Inflorescence a terminal truss of (8-)10-15(-20) flowers; rachis up to 5 mm long, usually less. Pedicels 15-20 mm long, stipitate-glandular. Calyx an undulate rim 2-3 mm long or with rounded glandular-ciliate lobes. Corolla white to rose-pink, usually variously flecked or spotted with crimson, sometimes heavily so, and blotched, sometimes unspotted, campanulate to tubular-campanulate, 40-50 mm long, 5(-7)-lobed with nectar-pouches. Stamens 10(-14), unequal, shorter than the corolla; filaments minutely pubescent in the lower half. Ovary stipitate-glandular and with sparse rufous dendroid hairs, with lobulate, glabrous basal disk. Style equalling corolla in length, glandular almost to tip. Stigma discoid, lobulate. Capsule up to about 3 cm long, curved.

Distribution. China (west Yunnan).

Habitat. In open forests and thickets or on their margins and on rocky slopes, 1800-3400 m.

Rhododendron × agastum
LILIAN SNELLING

RHODODENDRON × BATEMANII

Subgenus *Hymenanthes*
Section *Ponticum*? subsection *Campanulata* × subsection *Arborea*

When it was first discovered by Booth in the mountains of Bhutan, this hybrid was thought to be a new species and was described as such by J.D. Hooker. His formal description was published with Fitch's painting (tab. 5387) in *Curtis's Botanical Magazine* in 1863. Hooker named the plant after James Bateman of Knypersley Hall, Staffordshire, who raised it from seed sent by Booth to his kinsman, Nuttall, of Nutgrove, Cheshire, whence Bateman obtained his seed. Bateman presented a flowering plant to Kew, where the illustration was drawn, after which the plant was established in the then new 'Winter Garden'.

Although it is undoubtedly a hybrid, the parentage of *R.* × *batemanii* is not so certain. Chamberlain (1982) thought it might have arisen as a chance hybrid between *R. arboreum* Sm. *sensu lato* and *R. wallichii* Hook. fil. (or, more doubtfully, *R. campanulatum* D. Don) but pointed out that if this is the case, then the leaves, which can achieve 20 cm in length, are suspiciously large. He therefore included *R.* × *batemanii* only tentatively as a synonym of the above cross. Cox (1979) disagreed; he considered, as did Davidian (1989), that the plant is almost certainly a natural hybrid between *R. campanulatum* and *R. arboreum*, while doubt is cast by several authorities on the origin of Booth's seed.

Whatever its parentage may be, *R.* × *batemanii* is a fine plant, but at 6 m tall, is a little too enthusiastic for many gardens. The ochraceous to rusty brown soft indumentum of the lower leaf surface has an appeal of its own, but together with the large compact trusses of individually large, deep crimson to plum- purple flowers, it makes a splendid spectacle and the accompanying plate is a masterly example of Fitch's handiwork. For the latter reason the apparently otherwise undocumented *R.* × *batemanii* has been included here.

? **Rhododendron wallichii** Hook. fil. × **Rhododendron arboreum** J. Sm. *sens. lat.*, Chamberlain in Notes Roy. Bot. Gard. Edinburgh 39: 374 (1982).
? *R. batemanii* Hook. fil. in Bot. Mag. 89: t. 5387 (1863).
? *R. campanulatum* D. Don var. *campbellii* Millais, Rhododendrons ed. 1: 134 (1917).

Description. An evergreen, robust shrub up to 6 m or more in height with tomentose branchlets. Leaves coriaceous, elliptic or lanceolate-oblong, 10-20 cm long, up to about 5 cm wide, with obtuse or cuneate base, subacute apex and recurved margins, upper surface glabrous, lower surface with an ochreous or rusty brown soft tomentum. Petioles short, stout, tomentose. Inflorescence a compact truss of 12-20 large flowers. Pedicels about 2 cm long, pubescent. Calyx small, cupular, minutely and unequally 5-lobed. Corolla deep crimson to plum-purple with a few darker splashes on the uppermost lobes, more or less widely campanulate, about 6.5 cm across, the 5 lobes spreading and somewhat crumpled. Stamens 10, unequal; filaments pubescent in lower half. Ovary tomentose. Style slender, glabrous. Stigma small, discoid, lobulate.

Distribution. Doubtfully recorded from Bhutan without note of habitat. As the altitudinal ranges of *R. arboreum*, *R. campanulatum* and *R. wallichii* all overlap, the likely range for *R.* × *batemanii* could be anything from 1500 to 4500 m.

Rhododendron × batemanii
WALTER HOOD FITCH

RHODODENDRON × CLIVIANUM

Subgenus *Hymenanthes*
Section *Ponticum* subsection *Pontica* × subsection *Arborea*

This old hybrid, believed to be between *R. catawbiense* and a white form of *R. arboreum* is only rarely mentioned today and even more rarely, if ever, seen in gardens. The large, pink-flushed, white flowers with crimson to purple spots, produced in April and May are a little too early for the vagaries of the British climate and frequently incur damage from late frosts. Otherwise the plant is completely hardy, so that while the plant is young, it can be grown in a container outside, and brought into the glasshouse or conservatory to flower. Being strong-growing and adaptable, this rhododendron also makes a suitable subject for shaping into a standard, in which form it is ideally suited to glasshouse conditions. Eventually, of course, with such parentage, it is bound to become too large for such treatment, and will have to be planted out to take its chances of frost damage.

The hybrid was raised by Mr Iveson, the head gardener at Syon at the time, and exhibited in 1849 at an RHS show. It was named by Lindley, after the Dowager Duchess of Northumberland then residing at Syon House, and was figured in *Curtis's Botanical Magazine* (tab. 4478) as 'an example of what we are disposed to consider the most delicate and beautiful kind of Rhododendron yet in cultivation'. Thus wrote John Smith in the accompanying text, but he added no further descriptive notes, and I have been unable to find any additional information in relevant literature. The plate must therefore stand on its own merits, as a testament both to Fitch's expertise and a beautiful rhododendron which is currently and regrettably out of fashion.

Rhododendron catawbiense Michaux × **R. arboreum** Sm. *sens lat.*
R. clivianum J. Sm. in Bot. Mag. 75: t. 4478 (1849).

Rhododendron × clivianum
WALTER HOOD FITCH

GLOSSARY

accrescent: continuing to increase in size as the organism ages.
bistrate: two layers (in *Rhododendron*, usually referring to indumentum, especially of the leaf-surface; cf. unistrate).
bracteoles: floral bracts.
bullate: strongly puckered or wrinkled.
caducous: deciduous, referring, in *Rhododendron*, to organs other than leaves (e.g. perulae, scales).
clone: progeny produced vegetatively from one individual.
contiguous: adjoining, touching.
coriaceous: leathery.
dendroid: tree-like hairs, with a thick stalk or 'trunk'; further divided into folioliferous, flagellate, fasciculate, capitate, ramiform and cup-shaped types.
elepidote: without scales.
endemic: confined to a stated region.
fusiform: spindle-shaped.
glaucous: greyish, referring to the 'bloom' on leaves, etc.
grex: all hybrid progeny resulting from a single cross.
inarching: 'grafting by approach, the scion remaining partly attached to its parent, until union has taken place'.
indumentum: a covering, most frequently referring to leaves and stems.
lepidote: scaly.
loriform: multicellular hairs/bristles, on leaf-margins etc., often deciduous, but usually leaving persistent hair-bases.
perulae: vegetative bud-scales — also 'perules'.
precocious: flowers expanding before the leaves.
punctate: with minute dots.
scales: (in *Rhododendron*) minute dots on leaves, stems etc.
sessile: without a stem/stalk.
setae: stiff hairs/bristles, sometimes gland-tipped, sometimes flattened.
stellate: hairs with short to long, stiff, spreading apical branches.
stipitate: with a stem, usually referring to glands or scales.
stoloniferous: producing 'suckers'.
strigose: bristly.
taxa: plural of taxon, q.v.
taxon: a unit of classification of unspecified rank.
unistrate: one layer (in *Rhododendron*, usually referring to indumentum, especially of the leaf-surface; cf. bistrate).
ventricose: unequally swollen.
vesicular: bubble- or blister-like hairs.

APPENDIX

Rhododendron species that have appeared in *Curtis's Botanical Magazine*, 1787-2000

Correct name of species	Name under which published	Plate No.	Year	Artist/Engraver
adenogynum	adenogynum	9253	1929	L.Snelling
aequabile	aequabile	673	1974	M.Stones
afghanicum	afghanicum	8907	{1921}	L.Snelling
× agastum	agastum	9577	1939	L.Snelling
albiflorum	albiflorum	3670	1839	W.H.Fitch
albrechtii	albrechtii	9207	1928	L.Snelling
album	album	4972	1857	W.H.Fitch
'Altaclerense'	arboreum Altaclerense	3423	1835	J.Swan
amagianum	amagianum	379	1962	M.Stones
ambiguum	ambiguum	8400	1911	M.Smith
amesiae	amesiae	9221	1928	L.Snelling
annae	laxiflorum	385	1962	L.Snelling
anthopogon	anthopogon	3947	1842	W.H.Fitch
anthosphaerum	anthosphaerum	9083	1925	L.Snelling
aperantum	aperantum	9507	1937	L.Snelling
arboreum	arboreum limbatum	5311	1862	W.H.Fitch
arboreum	arboreum album	3290	1834	W.J.Hooker
arboreum	arboreum kingianum	7696	1900	M.Smith
arboreum ssp. arboreum	windsorii	5008	1857	W.H.Fitch
arboreum ssp. delavayi v. delavayi	delavayi	8137	1907	M.Smith
arboreum v. nilagiricum	nilagiricum	4381	1848	W.H.Fitch
arboreum v. nilagiricum	nilagiricum	9323	1933	L.Snelling
argyrophyllum ssp. hypoglaucum	hypoglaucum	8649	1916	M.Smith
argyrophyllum ssp. nankingense	argyrophyllum leiandrum	8767	1918	M.Smith
augustinii	augustinii	8497	1913	M.Smith
augustinii ssp. chasmanthum	chasmanthum	79	1949	L.Snelling
aureum	aureum	8882	{1921}	L.Snelling
auriculatum	auriculatum	8786	1919	M.Smith
aurigeranum	aurigeranum	787	1980	M.Stones
baileyi	baileyi	8942	1922	L.Snelling
balfourianum	balfourianum aganniphoides	531	1969	L.Snelling
× batemannii	batemani	5387	1863	W.H.Fitch
beanianum	beanianum	219	1953	A.V.Webster
beesianum	beesianum	125	1950	L.Snelling
boothii	boothii	7149	1890	M.Smith
brachyanthum	brachyanthum	8750	1918	M.Smith
brachyanthum ssp. hypolepidotum	hypolepidotum	9259	1929	L.Snelling
brachycarpum	brachycarpum	7881	1903	M.Smith
bracteatum	bracteatum	9031	1924	L.Snelling
× burmanicum	burmanicum	546	1969	M.Stones
burttii	burttii	130	1989	W.Walsh
calendulaceum	calendulaceum	3439	1835	J.Swan
callimorphum	callimorphum	8789	1919	M.Smith
calophytum v. calophytum	calophytum	9173	1927	L.Snelling
calostrotum ssp. calostrotum	calostrotum	9001	1923	L.Snelling
camelliiflorum	camelliaeflorum	4932	1856	W.H.Fitch

Correct name of species	Name under which published	Plate No.	Year	Artist/Engraver
campanulatum ssp. campanulatum	campanulatum	3759	1840	W.H.Fitch
campanulatum ssp. campanulatum	campanulatum	4928	1856	W.H.Fitch
campylocarpum ssp. campylocarpum	campylocarpum	4968	1857	W.H.Fitch
campylogynum	campylogynum	9407	1935	L.Snelling
camtschaticum	kamtschaticum	8210	1908	M.Smith
canadense	Rhodora canadensis	474	1800	S.Edwards
carneum	carneum	8634	1915	M.Smith
catawbiense	catawbiense	1671	1814	S.Edwards
caucasicum	caucaseum	1145	1808	S.Edwards
caucasicum	caucasicum straminea	3422	1835	W.J.Hooker
caucasicum × luteum (white)	caucasicum hybridum	3811	1841	W.H.Fitch
cerasinum	cerasinum	9538	1938	L.Snelling
championae	championae	4609	1851	W.H.Fitch
charitopes ssp. charitopes	charitopes	9358	1934	L.Snelling
christii	christi	29	1985	C.F.King
chrysodoron	chrysodoron	9442	1936	L.Snelling
ciliatum	ciliatum roseo-album	4648	1852	W.H.Fitch
ciliatum	modestum	7686	1899	M.Smith
ciliatum × glaucophyllum	wilsoni (hybridum)	5116	1859	W.H.Fitch
ciliicalyx	ciliicalyx	7782	1901	M.Smith
cinnabarinum	cinnabarinum pallidum	4788	1854	W.H.Fitch
cinnabarinum ssp. cinnabarinum	blandfordiaeflorum	4930	1856	W.H.Fitch
cinnabarinum ssp. concatenans	concatenans	634	1973	M.Stones
cinnamomeum	arboreum cinnamomeum	3825	1841	W.H.Fitch
citrinum v. citrinum	citrinum	4797	1854	W.H.Fitch
clementinae	clementinae	9392	1935	L.Snelling
× clivianum	clivianum	4478	1849	W.H.Fitch
collettianum	collettianum	7019	1888	M.Smith
concinnum	concinnum laetevirens	8912	{1921}	L.Snelling
concinnum	concinnum	8620	1915	M.Smith
concinnum	coombense	8280	1909	M.Smith
coriaceum	coriaceum	462	1965	S.Ross-Craig
crinigerum v. crinigerum	crinigerum	9464	1936	L.Snelling
cuffeanum	cuffeanum	8721	1917	M.Smith
culminicolum v. angiense	culminicolum angiense	268	1995	M.Fothergill
cuneatum	ravum	9561	1939	L.Snelling
cyanocarpum	cyanocarpum	155	1951	L.Snelling
dalhousiae v. dalhousiae	dalhousiae	73	1987	S.Wickison
dalhousiae v. dalhousiae	dalhousiae	4718	1853	W.H.Fitch
dalhousiae v. rhabdotum	rhabdotum	9447	1936	L.Snelling
dauricum	dahuricum sempervirens	1888	1817	Weddell
dauricum	dauricum sempervirens	8930	{1921}	L.Snelling
dauricum	dauricum	636	1803	S.Edwards
davidsonianum	charianthum	8665	1916	M.Smith
davidsonianum	davidsonianum	8605	1915	M.Smith
decorum	decorum	8659	1916	M.Smith
decorum ssp. diaprepes	diaprepes	9524	1938	L.Snelling
degronianum ssp. degronianum	japonicum pentamerum	8403	1911	M.Smith
dendricola	dendricola	9682	1948	L.Snelling
dendricola	taronense	1	1948	L.Snelling
× detonsum	detonsum	9359	1934	L.Snelling

Correct name of species	Name under which published	Plate No.	Year	Artist/Engraver
dichroanthum ssp. apodectum	apodectum	9014	1923	L.Snelling
dichroanthum ssp. dichroanthum	dichroanthum	8815	1919	M.Smith
dielsianum	dielsianum	16	1984	C.F.King
dilatatum	dilatatum	7681	1899	M.Smith
edgeworthii	edgeworthii	4936	1856	W.H.Fitch
edgeworthii × ciliatum 'Princess Alice'	'Princess Alice'	191	1992	W.H.Fitch
elliottii	elliottii	9546	1938	L.Snelling
faberi ssp. prattii	prattii	9414	1935	L.Snelling
facetum	eriogynum	9337	1934	L.Snelling
falconeri ssp. eximium	falconeri eximia	7317	1893	M.Smith
falconeri ssp. falconeri	falconeri	4924	1856	W.H.Fitch
flavidum	flavidum	8326	1910	M.Smith
fletcherianum	fletcherianum	508	1967	L.Snelling
floccigerum	floccigerum	9290	1929	L.Snelling
floribundum	floribundum	9609	1940	L.Snelling
formosum v. formosum	formosum	4457	1849	W.H.Fitch
formosum v. formosum	iteophyllum	563	1970	M.Stones
formosum v. inaequale	inaequale	295	1957	A.V.Webster
formosum × dalhousiae v. dalhousiae	dalhousiae hybridum	5322	1862	W.H.Fitch
forrestii ssp. forrestii	forrestii	9186	1927	L.Snelling
fortunei ssp. discolor	discolor	8696	1917	M.Smith
fortunei ssp. discolor	discolor	55	1986	V.Price
fortunei ssp. fortunei	fortunei	5596	1866	W.H.Fitch
fulgens	fulgens	5317	1862	W.H.Fitch
fulvum	fulvum	9587	1939	L.Snelling
galactinum	galactinum	231	1954	L.Snelling
genesterianum	genesterianum	9310	1933	L.Snelling
glaucophyllum v. glaucophyllum	glaucum	4721	1853	W.H.Fitch
glischrum ssp. glischrum	glischrum	9035	1924	L.Snelling
goodenoughii	goodenoughii	826	1981	M.Stones
grande	argenteum	5054	1858	W.H.Fitch
grande	grande roseum	6948	1887	M.Smith
griersonianum	griersonianum	9195	1927	L.Snelling
griffithianum	griffithianum aucklandii	5065	1858	W.H.Fitch
haematodes ssp. chaetomallum	chaetomallum	25	1948	L.Snelling
haematodes ssp. haematodes	haematodes	9165	1927	L.Snelling
hanceanum	hanceanum	8669	1916	M.Smith
heliolepis v. brevistylum	brevistylum	8898	{1921}	L.Snelling
hippophaeoides	hippophaeoides	9156	1926	L.Snelling
hirsutum	hirsutum	1853	1816	Weddell
hodgsonii	hodgsoni	5552	1866	W.H.Fitch
hookeri	hookeri	4926	1856	W.H.Fitch
hyperythrum	hyperythrum	109	1950	L.Snelling
impeditum	impeditum	489	1966	L.Snelling
impositum	impositum	816	1981	C.F.King
insigne	insigne	8885	{1921}	L.Snelling
intricatum	intricatum	8163	1907	M.Smith
irroratum ssp. irroratum	irroratum	7361	1894	M.Smith
jasminiflorum	jasminiflorum	4524	1850	W.H.Fitch
javanicum	javanicum	4336	1847	W.H.Fitch
javanicum ssp. brookeanum	brookeanum	4935	1856	W.H.Fitch

Correct name of species	Name under which published	Plate No.	Year	Artist/Engraver
javanicum ssp. brookeanum	brookeanum	726	1977	M.Stones
javanicum ssp. brookeanum	brookeanum	192	1992	W.H.Fitch
keiskei	keiskei	8300	1910	M.Smith
kendrickii	kendrickii latifolium	5129	1859	W.H.Fitch
kendrickii	shepherdii	5125	1859	W.H.Fitch
keysii	keysii	4875	1855	W.H.Fitch
kiusianum 'Amoenum'	amoena	4728	1853	W.H.Fitch
kongboense	kongboense	9492	1937	L.Snelling
konori	konori	107	1988	R.Purves
kyawi	kyawii	9271	1929	L.Snelling
lacteum	lacteum	8988	1923	L.Snelling
lapponicum	lapponicum	3106	1831	J.Swan
lapponicum	parvifolium	9229	1928	L.Snelling
lepidotum	lepidotum obovatum	6450	1879	H.Thiselton-Dyer
lepidotum	lepidotum	4657	1852	W.H.Fitch
lepidotum	lepidotum	4802	1854	W.H.Fitch
leptothrium	leptothrium	502	1967	L.Snelling
leucaspis	leucaspis	9665	1944	S.Ross-Craig
lindleyi	lindleyi	363	1960	M.Stones
lineare	lineare	103	1988	R.Purves
lochae	lochae	9651	1943	S.Ross-Craig
longesquamatum	longesquamatum	9430	1936	L.Snelling
longiflorum	javanicum tubiflora	6850	1885	M.Smith
ludlowii	ludlowii	412	1963	M.Stones
ludwigianum	ludwigianum	748	1978	M.Stones
lutescens	lutescens	8851	1920	M.Smith
lyi	lyi	9051	1924	L.Snelling
macabeanum	macabeanum	187	1952	S.Ross-Craig
macgregoriae	macgregoriae	552	1969	M.Stones
macrophyllum	californicum	4863	1855	W.H.Fitch
maddenii ssp. crassum	crassum	9673	1946	L.Snelling
maddenii ssp. crassum	maddeni obtusifolia	8212	1908	M.Smith
maddenii ssp. maddenii	calophyllum	5002	1857	W.H.Fitch
maddenii ssp. maddenii	maddeni	4805	1854	W.H.Fitch
maddenii ssp. maddenii	meddianum	9636	1942	L.Snelling
malayanum	malayanum	6045	1873	W.H.Fitch
mallotum	mallotum	9419	1935	L.Snelling
mariesii	mariesii	8206	1908	M.Smith
maximum	maximum	951	1806	S.Edwards
maximum	maximum	3454	1835	J.Swan
megacalyx	megacalyx	9326	1933	L.Snelling
megeratum	megeratum	9120	1926	L.Snelling
mekongense v. melinanthum	melinanthum	8903	{1921}	L.Snelling
micranthum	micranthum	8198	1908	M.Smith
minus v. minus	punctatum carolina	2285	1822	J.Curtis
molle	sinense	5905	1871	W.H.Fitch
monanthum	sulfureum	8946	1922	L.Snelling
morii	morii	517	1968	L.Snelling
moulmainense	moulmainense	4904	1856	W.H.Fitch
moulmainense	stenaulum	9656	1944	L.Snelling
moupinense	moupinense	8598	1915	M.Smith

Correct name of species	Name under which published	Plate No.	Year	Artist/Engraver
mucronulatum	mucronulatum	8304	1910	M.Smith
multicolor	multicolor	6769	1884	M.Smith
myrtifolium	kotschyi	9132	1926	L.Snelling
neriiflorum ssp. neriiflorum	neriiflorum	8727	1917	M.Smith
neriiflorum ssp. neriiflorum	phoenicodon	9521	1938	L.Snelling
nipponicum	nipponicum	491	1966	M.Stones
nivale ssp. boreale	nigropunctatum	8529	1913	M.Smith
niveum	niveum fulva	6827	1885	M.Smith
niveum	niveum	4730	1853	W.H.Fitch
nuttallii	nuttallii	5146	1859	W.H.Fitch
oldhamii	oldhamii	9059	1924	L.Snelling
orbiculare	orbiculare	8775	1918	M.Smith
orbiculatum	orbiculatum	575	1970	M.Stones
oreodoxa v. fargesii	erubescens	8643	1916	M.Smith
oreodoxa v. fargesii	fargesii	8736	1917	M.Smith
oreodoxa v. oreodoxa	haematocheilum	8518	1913	M.Smith
oreotrephes	oreotrephes	8784	1918	M.Smith
orthocladum v. microleucum	microleucum	171A	1951	L.Snelling
ovatum	bachii	9375	1934	L.Snelling
pachypodum	scottianum	9238	1928	L.Snelling
pachytrichum	monosematum	8675	1916	M.Smith
parmulatum	parmulatum	9624	1941	L.Snelling
periclymenoides	nudiflorum scintillans	3667	1839	J.Swan
phaeochitum	phaeochitum	766	1978	M.Stones
planetum	planetum	8953	1922	L.Snelling
pleianthum	pleianthum	88	1987	C.F.King
polylepis	harrovianum	8309	1910	M.Smith
ponticum	ponticum	650	1803	S.Edwards
praetervisum	praetervisum	104	1988	R.Purves
principis	vellereum	147	1951	L.Snelling
protistum v. giganteum	giganteum	253	1955	A.V.Webster
pseudochrysanthum	pseudochrysanthum	284	1956	L.Snelling
pubescens	pubescens	9319	1933	L.Snelling
racemosum	racemosum	7301	1893	M.Smith
rarilepidotum	rarilepidotum	6	1984	C.F.King
rarum	rarum	610	1972	M.Stones
reticulatum	rhombicum	6972	1887	M.Smith
retusum	retusum	4859	1855	W.H.Fitch
roxieanum	roxieanum	9383	1935	L.Snelling
rubiginosum	desquamatum	9497	1937	L.Snelling
rubiginosum	rubiginosum	7621	1898	M.Smith
rugosum	rugosum	808	1981	C.F.King
rupicola v. chryseum	chryseum	9246	1928	L.Snelling
russatum	cantabile	8963	1922	L.Snelling
russatum	russatum	161	1990	V.Price
salicifolium	salicifolium	106	1988	R.Purves
saluenense ssp. chamaeunum	prostratum	8747	1918	M.Smith
saluenense ssp. saluenense	saluenense	9095	1925	L.Snelling
sanguineum ssp. didymum	didymum	9217	1928	W.Trevithick/L.Snelling
sanguineum ssp. sang. var. haemaleum	sanguineum	9263	1929	L.Snelling
sargentianum	sargentianum	8871	1920	M.Smith

Correct name of species	Name under which published	Plate No.	Year	Artist/Engraver
scabrifolium	scabrifolium	7159	1891	M.Smith
scabrum ssp. scabrum	sublanceolatum	8478	1913	M.Smith
schlippenbachii	schlippenbachii	7373	1894	M.Smith
scopulorum	scopulorum	9399	1935	L.Snelling
searsiae	searsiae	8993	1923	L.Snelling
semibarbatum	semibarbatum	9147	1926	L.Snelling
semnoides	semnoides	18	1948	L.Snelling
serotinum	serotinum	8841	1920	M.Smith
serpyllifolium	serpyllifolium	7503	1896	M.Smith
setosum	setosum	8523	1913	M.Smith
sherriffii	sherriffii	337	1959	M.Stones
sidereum	sidereum	638	1973	L.Snelling
siderophyllum	siderophyllum	8759	1918	M.Smith
sikangense v. exquisitum	exquisitum	9597	1939	S.Ross-Craig
simiarum	fordii	8111	1906	M.Smith
sinogrande	sinogrande	8973	1922	L.Snelling
smirnovii	smirnovi	7495	1896	M.Smith
smithii	smithii	5120	1859	W.H.Fitch
souliei	souliei	8622	1915	M.Smith
sperabile v. sperabile	sperabile	9301	1933	L.Snelling
spinuliferum	spinuliferum	8408	1911	M.Smith
stamineum	stamineum	8601	1915	M.Smith
stapfianum	lacteum	8372	1911	M.Smith
stenophyllum	stenophyllum	105	1988	R.Purves
strigillosum	strigillosum	8864	1920	M.Smith
suaveolens	suaveolens	600	1972	M.Stones
superbum	superbum	108	1988	R.Purves
sutchuenense	sutchuenense	8362	1911	M.Smith
taggianum	headfortianum	9614	1942	L.Snelling
taggianum	taggianum	9612	1942	L.Snelling
tephropeplum	tephropeplum	9343	1934	L.Snelling
thayerianum	thayerianum	8983	1923	L.Snelling
thomsonii ssp. thomsonii	thomsoni	4997	1857	W.H.Fitch
tosaense	tosaense	52	1949	L.Snelling
traillianum v. traillianum	traillianum	8900	{1921}	L.Snelling
trichanthum	villosum	8880	{1921}	L.Snelling
trichocladum	trichocladum	9073	1925	L.Snelling
trichostomum	hedyosmum	9202	1927	L.Snelling
trichostomum	ledoides	8831	1920	M.Smith
ungernii	ungernii	8332	1910	M.Smith
uniflorum v. imperator	imperator	514	1967	L.Snelling
uvarifolium	niphargum	9480	1937	L.Snelling
valentinianum	valentinianum	623	1972	M.Stones
vaseyi	vaseyi	8081	1906	M.Smith
veitchianum	cubittii	9502	1937	L.Snelling
veitchianum	veitchianum	4992	1857	W.H.Fitch
vernicosum	vernicosum	8834	1920	M.Smith
vernicosum	vernicosum	8904	{1921}	L.Snelling
vernicosum	vernicosum	8905	{1921}	L.Snelling
virgatum ssp. oleifolium	oleifolium	8802	1919	M.Smith
virgatum ssp. virgatum	virgatum	5060	1858	W.H.Fitch

Correct name of species	Name under which published	Plate No.	Year	Artist/Engraver
wardii v. wardii	wardii	587	1972	M.Stones
wasonii	wasonii	9190	1927	L.Snelling
weyrichii	weyrichii	9475	1937	L.Snelling
wightii	wightii	8492	1913	M.Smith
williamsianum	williamsianum	8935	1922	L.Snelling
wiltonii	wiltonii	9388	1935	L.Snelling
wrightianum	wrightianum	653	1973	M.Stones
yakushimanum	yakushimanum	771	1979	M.Stones
yedoense v. poukhanense	yedoense poukhanense	455	1964	M.Stones
yunnanense	yunnanense	7614	1898	M.Smith
zaleucum	zaleucum	8878	{1921}	L.Snelling
zoelleri	zoelleri	682	1975	M.Stones

REFERENCES AND SELECTED BIBLIOGRAPHY

Bean, W.J. (1976). *Trees and Shrubs Hardy in the British Isles*, ed. 8: 537. John Murray, London.

Bean, W.J. (1988). *Trees and Shrubs Hardy in the British Isles*, ed. 8, Supplement: 419. John Murray, London.

Chamberlain, D.F. (1982). A Revision of Rhododendron 2, subgenus Hymenanthes. In: *Notes from the Royal Botanic Gardens Edinburgh* 39(2).

Chamberlain, D.F. & Rae, S.T. (1990). A Revision of Rhododendron 4, subgenus Tsutsusi. In: *Edinburgh Journal of Botany* 47(2).

Chamberlain, D. *et al.* (1996). *The Genus Rhododendron. Its classification & synonymy*. Royal Botanic Garden, Edinburgh.

Cox, P.A. (1973). *Dwarf Rhododendrons*. Batsford, London.

Cox, P.A. (1985). *The Smaller Rhododendrons*. Batsford, London.

Cox, P.A. (1990). *The Larger Rhododendrons* (revised edition). Batsford, London.

Cullen, J. (1980). A Revision of Rhododendron 1, subgenus Rhododendron. In: *Notes from the Royal Botanic Gardens Edinburgh* 39(1).

Curtis, W. (1787-2000). *Curtis's Botanical Magazine* (incorporating *The Kew Magazine* 1984-1994).

Davidian, H.H. (1982). *The Rhododendron Species*, 1. Lepidotes. Batsford, London.

Davidian, H.H. (1989). *The Rhododendron Species*, 2. Elepidotes (Arboreum-Lacteum). Batsford, London.

Davidian, H.H. (1992). *The Rhododendron Species*, 3. Elepidotes (Thomsonii, Azaleastrum & Camschaticum). Batsford, London.

Hooker, J.D. (1849-1851). *Rhododendrons of the Sikkim-Himalaya.*

Jackson, B.D. (1928). *A Glossary of Botanic Terms*. Duckworth, London.

Kingdon Ward, F. (1924). *The Romance of Plant Hunting*. Arnold, London.

Kingdon Ward, F. (1949). *Rhododendrons*. Latimer House, London.

Kron, K.A. (1993). A Revision of Rhododendron section Pentanthera. In: *Edinburgh Journal of Botany* 50(3).

Millais, J.G. (1917 & 1924). *Rhododendrons and the Various Hybrids.*

Philipson, M.N. & Philipson, W.R. (1975). A Revision of Rhododendron section Lapponicum. In: *Notes from the Royal Botanic Gardens Edinburgh* 34(1).

Philipson, M.N. & Philipson, W.R. (1982). A Preliminary Synopsis of the Genus Rhododendron 3. In: *Notes from the Royal Botanic Gardens Edinburgh* 40(1).

Philipson, M.N. & Philipson, W.R. (1986). A Revision of Rhododendron 3, subgenera Azaleastrum, Mumeazalea, Candidastrum and Therorhodion. In: *Notes from the Royal Botanic Gardens Edinburgh* 44(1).

Royal Horticultural Society. (1980). *The Rhododendron Handbook* (revised periodically). RHS, London.

Stevenson, J.B. (ed.). (1930). *The Species of Rhododendron*. The Rhododendron Society, Ascot.

VIREYA REFERENCES

Argent, G. (1988). Some Attractive Vireya Rhododendrons. In: *Curtis's Botanical Magazine* (*Kew Magazine*) 5(3).

Argent, G. et al. (1989). *Rhododendrons of Sabah*. Sabah Parks Trustees, Kota Kinabalu.

Rouse, J.L. (1986). Raising Vireyas from Seed. In: *The Rhododendron* 26(1): 8.

Sinclair, I. (1984). A New Compost for Vireya Rhododendrons. In: *The Plantsman* 6(2): 101.

Sleumer, H. (1966). *Flora Malesiana* 6(4). Kluwer Academic Publishers, Dordrecht.

Smith, J.C. (1989). *Vireya Rhododendrons.*